2018 SQA Specimen and Past Papers with Answers

Higher
MATHEMATICS

2017 & 2018 Exams
and 2018 Specimen Question Paper

(Side tab: Higher MATHEMATICS)

(Side tab: SQA)

HODDER
GIBSON
AN HACHETTE UK COMPANY

This book contains the official SQA 2017 and 2018 Exams, and the 2018 Specimen Question Paper for Higher Maths, with associated SQA-approved answers modified from the official marking instructions that accompany the paper.

In addition the book contains study skills advice. This advice has been specially commissioned by Hodder Gibson, and has been written by experienced senior teachers and examiners in line with the Higher syllabus and assessment outlines. This is not SQA material but has been devised to provide further practice for Higher examinations.

Every effort has been made to trace the copyright holders and to obtain their permission for the use of copyright material. Hodder Gibson will be happy to receive information allowing us to rectify any error or omission in future editions.

Hachette UK's policy is to use papers that are natural, renewable and recyclable products and made from wood grown in sustainable forests. The logging and manufacturing processes are expected to conform to the environmental regulations of the country of origin.

Orders: please contact Bookpoint Ltd, 130 Park Drive, Milton Park, Abingdon, Oxon OX14 4SE. Telephone: (44) 01235 827827. Fax: (44) 01235 400454. Lines are open 9.00–5.00, Monday to Saturday, with a 24-hour message answering service. Visit our website at www.hoddereducation.co.uk. Hodder Gibson can also be contacted directly at hoddergibson@hodder.co.uk

This collection first published in 2018 by
Hodder Gibson, an imprint of Hodder Education,
An Hachette UK Company
211 St Vincent Street
Glasgow G2 5QY

Typeset by Aptara, Inc.

Printed in the UK

A catalogue record for this title is available from the British Library

ISBN: 978-1-5104-5675-4

2 1

2019 2018

Introduction

Higher Mathematics

This book of SQA past papers contains the question papers used in the 2017 and 2018 exams (with the answers at the back of the book). A specimen question paper reflecting the requirements, content and duration of the revised exam from 2019 is also included. All of the question papers included in the book provide excellent, representative practice for the final exams.

Using the 2017 and 2018 past papers as part of your revision will help you to develop the vital skills and techniques needed for the exam, and will help you to identify any knowledge gaps you may have.

It is always a very good idea to refer to SQA's website for the most up-to-date course specification documents. Further details can be found in the Higher Mathematics section on the SQA website: https://www.sqa.org.uk/sqa/45750.html

The course

The Higher Mathematics course aims to:

- motivate and challenge learners by enabling them to select and apply mathematical techniques in a variety of mathematical situations
- develop confidence in the subject and a positive attitude towards further study in mathematics and the use of mathematics in employment
- deliver in-depth study of mathematical concepts and the ways in which mathematics describes our world
- allow learners to interpret, communicate and manage information in mathematical form; skills which are vital to scientific and technological research and development
- deepen learners' skills in using mathematical language and exploring advanced mathematical ideas.

The Higher qualification in Mathematics is designed to build upon and extend learners' mathematical skills, knowledge and understanding in a way that recognises problem solving as an essential skill and enables them to integrate their knowledge of different aspects of the subject.

You will acquire an enhanced awareness of the importance of mathematics to technology and to society in general. Where appropriate, mathematics will be developed in context, and mathematical techniques will be applied in social and vocational contexts related to likely progression routes such as commerce, engineering and science where the mathematics learned will be put to direct use.

The syllabus is designed to build upon your prior learning in the areas of algebra, geometry and trigonometry and to introduce you to elementary calculus.

How the course is assessed

- To gain the course award, you must pass an examination.
- The examination is set and marked by the SQA.
- The course award is graded A–D, the grade being determined by the total mark you score in the examination.

The examination

- The examination consists of two papers. The number of marks and the times allotted for the papers are:

  Paper 1
  (non-calculator)   70 marks   1 hour 30 minutes

  Paper 2   80 marks   1 hour 45 minutes

- Both papers contain short and extended response questions in which candidates are required to apply numerical, algebraic, geometric, trigonometric, calculus, and reasoning skills.
- The examination is designed so that approximately 65% of the marks will be available for level C responses.
- Some questions will assess only operational skills (65% of the marks) but other questions will require both operational and reasoning skills (35% of the marks).

Further details can be found in the Higher Mathematics section on the SQA website: http://www.sqa.org.uk/sqa/47910.html.

Key tips for your success

Practise! Practise! Practise!

DOING maths questions is the most effective use of your study time. You will benefit much more from spending 30 minutes doing maths questions than spending several hours copying out notes or reading a maths textbook.

Basic skills

You must practise the following essential basic skills for Higher Mathematics throughout the duration of this course – expanding brackets; solving equations; manipulating algebraic expressions; and, in particular, working with exact values with trigonometric expressions and equations.

Non-routine problems

It is important to practise non-routine problems as often as possible throughout the course, particularly if you are aiming for an A or B pass in Higher Mathematics.

Graph sketching

Graph sketching is an important and integral part of Mathematics. Ensure that you practise sketching graphs on plain paper whenever possible throughout this course. Neither squared nor graph paper are allowed in the Higher Mathematics examination.

Marking instructions

Ensure that you look at the detailed marking instructions of past papers. They provide further advice and guidelines as well as showing you precisely where, and for what, marks are awarded.

Show all working clearly

The instructions on the front of the exam paper state that *"Full credit will be given only to solutions which contain appropriate working."* A "correct" answer with no working may only be awarded partial marks or even no marks at all. An incomplete answer will be awarded marks for any appropriate working. Attempt every question, even if you are not sure whether you are correct or not. Your solution may contain working which will gain some marks. A blank response is certain to be awarded no marks. Never score out working unless you have something better to replace it with.

Make drawings

Try drawing what you visualise as the "picture", described within the wording of each relevant question. This is a mathematical skill expected of most candidates at Higher level. Making a rough sketch of the diagram in your answer booklet may also help you interpret the question and achieve more marks.

Extended response questions

You should look for connections between parts of questions, particularly where there are three or four sections to a question. These are almost always linked and, in some instances, an earlier result in part (a) or (b) is needed and its use would avoid further repeated work.

Notation

In all questions, make sure that you use the correct notation. In particular, for integration questions, remember to include 'dx' within your integral.

Radians

Remember to work in radians when attempting any calculus questions involving trigonometric functions.

Simplify

Get into the habit of simplifying expressions before doing any further work with them. This should make all subsequent work easier.

Subtraction

Be careful when subtracting one expression from another: ensure that any negative is applied correctly.

Good luck!

Remember that the rewards for passing Higher Mathematics are well worth it! Your pass will help you get the future you want for yourself. In the exam, be confident in your own ability. If you're not sure how to answer a question, trust your instincts and just give it a go anyway – keep calm and don't panic! GOOD LUCK!

Study Skills – what you need to know to pass exams!

General exam revision: 20 top tips

When preparing for exams, it is easy to feel unsure of where to start or how to revise. This guide to general exam revision provides a good starting place, and, as these are very general tips, they can be applied to all your exams.

1. Start revising in good time.

Don't leave revision until the last minute – this will make you panic and it will be difficult to learn. Make a revision timetable that counts down the weeks to go.

2. Work to a study plan.

Set up sessions of work spread through the weeks ahead. Make sure each session has a focus and a clear purpose. What will you study, when and why? Be realistic about what you can achieve in each session, and don't be afraid to adjust your plans as needed.

3. Make sure you know exactly when your exams are.

Get your exam dates from the SQA website and use the timetable builder tool to create your own exam schedule. You will also get a personalised timetable from your school, but this might not be until close to the exam period.

4. Make sure that you know the topics that make up each course.

Studying is easier if material is in manageable chunks – why not use the SQA topic headings or create your own from your class notes? Ask your teacher for help on this if you are not sure.

5. Break the chunks up into even smaller bits.

The small chunks should be easier to cope with. Remember that they fit together to make larger ideas. Even the process of chunking down will help!

6. Ask yourself these key questions for each course:

- Are all topics compulsory or are there choices?
- Which topics seem to come up time and time again?
- Which topics are your strongest and which are your weakest?

Use your answers to these questions to work out how much time you will need to spend revising each topic.

7. Make sure you know what to expect in the exam.

The subject-specific introduction to this book will help with this. Make sure you can answer these questions:

- How is the paper structured?
- How much time is there for each part of the exam?
- What types of question are involved? These will vary depending on the subject so read the subject-specific section carefully.

8. Past papers are a vital revision tool!

Use past papers to support your revision wherever possible. This book contains the answers and mark schemes too – refer to these carefully when checking your work. Using the mark scheme is useful; even if you don't manage to get all the marks available first time when you first practise, it helps you identify how to extend and develop your answers to get more marks next time – and of course, in the real exam.

9. Use study methods that work well for you.

People study and learn in different ways. Reading and looking at diagrams suits some students. Others prefer to listen and hear material – what about reading out loud or getting a friend or family member to do this for you? You could also record and play back material.

10. There are three tried and tested ways to make material stick in your long-term memory:

- Practising – e.g. rehearsal, repeating
- Organising – e.g. making drawings, lists, diagrams, tables, memory aids
- Elaborating – e.g. incorporating the material into a story or an imagined journey

11. Learn actively.

Most people prefer to learn actively – for example, making notes, highlighting, redrawing and redrafting, making up memory aids, or writing past paper answers. A good way to stay engaged and inspired is to mix and match these methods – find the combination that best suits you. This is likely to vary depending on the topic or subject.

12. Be an expert.

Be sure to have a few areas in which you feel you are an expert. This often works because at least some of them will come up, which can boost confidence.

13. Try some visual methods.

Use symbols, diagrams, charts, flashcards, post-it notes etc. Don't forget – the brain takes in chunked images more easily than loads of text.

14. Remember – practice makes perfect.

Work on difficult areas again and again. Look and read – then test yourself. You cannot do this too much.

15. Try past papers against the clock.

Practise writing answers in a set time. This is a good habit from the start but is especially important when you get closer to exam time.

16. Collaborate with friends.

Test each other and talk about the material – this can really help. Two brains are better than one! It is amazing how talking about a problem can help you solve it.

17. Know your weaknesses.

Ask your teacher for help to identify what you don't know. Try to do this as early as possible. If you are having trouble, it is probably with a difficult topic, so your teacher will already be aware of this – most students will find it tough.

18. Have your materials organised and ready.

Know what is needed for each exam:

- Do you need a calculator or a ruler?
- Should you have pencils as well as pens?
- Will you need water or paper tissues?

19. Make full use of school resources.

Find out what support is on offer:

- Are there study classes available?
- When is the library open?
- When is the best time to ask for extra help?
- Can you borrow textbooks, study guides, past papers, etc.?
- Is school open for Easter revision?

20. Keep fit and healthy!

Try to stick to a routine as much as possible, including with sleep. If you are tired, sluggish or dehydrated, it is difficult to see how concentration is even possible. Combine study with relaxation, drink plenty of water, eat sensibly, and get fresh air and exercise – all these things will help more than you could imagine. Good luck!

HIGHER

2017

National
Qualifications
2017

X747/76/11

Mathematics
Paper 1
(Non-Calculator)

FRIDAY, 5 MAY

9:00 AM – 10:10 AM

Total marks — 60

Attempt ALL questions.

You may NOT use a calculator.

Full credit will be given only to solutions which contain appropriate working.

State the units for your answer where appropriate.

Answers obtained by readings from scale drawings will not receive any credit.

Write your answers clearly in the spaces provided in the answer booklet. The size of the space provided for an answer should not be taken as an indication of how much to write. It is not necessary to use all the space.

Additional space for answers is provided at the end of the answer booklet. If you use this space **you must clearly identify the question number** you are attempting.

Use **blue** or **black** ink.

Before leaving the examination room you must give your answer booklet to the Invigilator; if you do not, you may lose all the marks for this paper.

FORMULAE LIST

Circle:

The equation $x^2 + y^2 + 2gx + 2fy + c = 0$ represents a circle centre $(-g, -f)$ and radius $\sqrt{g^2 + f^2 - c}$.

The equation $(x - a)^2 + (y - b)^2 = r^2$ represents a circle centre (a, b) and radius r.

Scalar Product: $\mathbf{a}.\mathbf{b} = |\mathbf{a}||\mathbf{b}| \cos \theta$, where θ is the angle between $\mathbf{a}$ and $\mathbf{b}$

or $\mathbf{a}.\mathbf{b} = a_1b_1 + a_2b_2 + a_3b_3$ where $\mathbf{a} = \begin{pmatrix} a_1 \\ a_2 \\ a_3 \end{pmatrix}$ and $\mathbf{b} = \begin{pmatrix} b_1 \\ b_2 \\ b_3 \end{pmatrix}$.

Trigonometric formulae:

$$\sin (A \pm B) = \sin A \cos B \pm \cos A \sin B$$
$$\cos (A \pm B) = \cos A \cos B \mp \sin A \sin B$$
$$\sin 2A = 2 \sin A \cos A$$
$$\cos 2A = \cos^2 A - \sin^2 A$$
$$= 2 \cos^2 A - 1$$
$$= 1 - 2 \sin^2 A$$

Table of standard derivatives:

$f(x)$	$f'(x)$
$\sin ax$	$a \cos ax$
$\cos ax$	$-a \sin ax$

Table of standard integrals:

$f(x)$	$\int f(x)dx$
$\sin ax$	$-\dfrac{1}{a} \cos ax + c$
$\cos ax$	$\dfrac{1}{a} \sin ax + c$

Page two

MARKS

Attempt ALL questions

Total marks — 60

1. Functions f and g are defined on suitable domains by $f(x) = 5x$ and $g(x) = 2\cos x$.

   (a) Evaluate $f(g(0))$.   1

   (b) Find an expression for $g(f(x))$.   2

2. The point $P(-2, 1)$ lies on the circle $x^2 + y^2 - 8x - 6y - 15 = 0$.

   Find the equation of the tangent to the circle at P.   4

3. Given $y = (4x-1)^{12}$, find $\dfrac{dy}{dx}$.   2

4. Find the value of k for which the equation $x^2 + 4x + (k-5) = 0$ has equal roots.   3

5. Vectors **u** and **v** are $\begin{pmatrix} 5 \\ 1 \\ -1 \end{pmatrix}$ and $\begin{pmatrix} 3 \\ -8 \\ 6 \end{pmatrix}$ respectively.

   (a) Evaluate **u.v**.   1

   (b)

   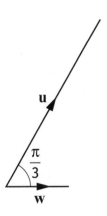

   Vector **w** makes an angle of $\dfrac{\pi}{3}$ with **u** and $|\mathbf{w}| = \sqrt{3}$.

   Calculate **u.w**.   3

MARKS

6. A function, h, is defined by $h(x) = x^3 + 7$, where $x \in \mathbb{R}$.

   Determine an expression for $h^{-1}(x)$.   3

7. A$(-3, 5)$, B$(7, 9)$ and C$(2, 11)$ are the vertices of a triangle.

   Find the equation of the median through C.   3

8. Calculate the rate of change of $d(t) = \dfrac{1}{2t}$, $t \neq 0$, when $t = 5$.   3

9. A sequence is generated by the recurrence relation $u_{n+1} = mu_n + 6$ where m is a constant.

   (a) Given $u_1 = 28$ and $u_2 = 13$, find the value of m.   2

   (b) (i) Explain why this sequence approaches a limit as $n \to \infty$.   1

        (ii) Calculate this limit.   2

MARKS

10. Two curves with equations $y = x^3 - 4x^2 + 3x + 1$ and $y = x^2 - 3x + 1$ intersect as shown in the diagram.

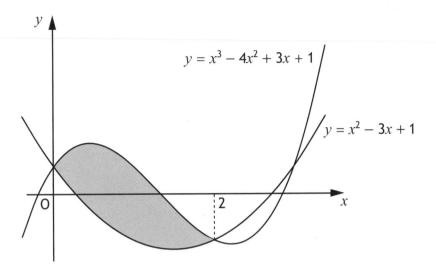

(a) Calculate the shaded area.   **5**

The line passing through the points of intersection of the curves has equation $y = 1 - x$.

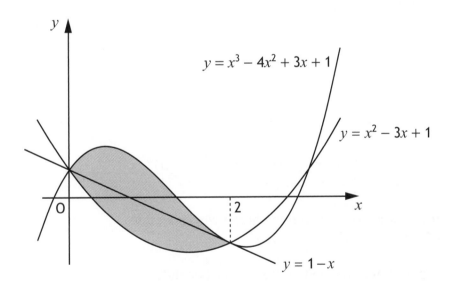

(b) Determine the fraction of the shaded area which lies below the line $y = 1 - x$.   **4**

[Turn over

MARKS

11. A and B are the points $(-7, 2)$ and $(5, a)$.

    AB is parallel to the line with equation $3y - 2x = 4$.

    Determine the value of a.   3

12. Given that $\log_a 36 - \log_a 4 = \dfrac{1}{2}$, find the value of a.   3

13. Find $\displaystyle\int \dfrac{1}{(5-4x)^{\frac{1}{2}}}\,dx, \quad x < \dfrac{5}{4}$.   4

14. (a) Express $\sqrt{3}\sin x° - \cos x°$ in the form $k\sin(x-a)°$,
       where $k > 0$ and $0 < a < 360$.   4

    (b) Hence, or otherwise, sketch the graph with equation
       $y = \sqrt{3}\sin x° - \cos x°$, $0 \le x \le 360$.   3

       Use the diagram provided in the answer booklet.

MARKS

15. A quadratic function, f, is defined on $\mathbb{R}$, the set of real numbers.

Diagram 1 shows part of the graph with equation $y = f(x)$.
The turning point is $(2, 3)$.

Diagram 2 shows part of the graph with equation $y = h(x)$.
The turning point is $(7, 6)$.

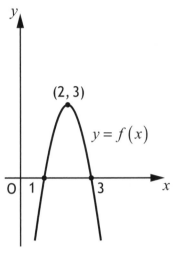 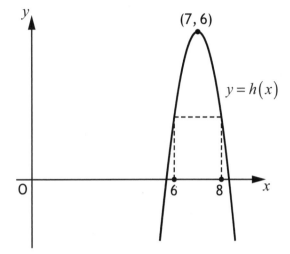

Diagram 1   Diagram 2

(a) Given that $h(x) = f(x+a) + b$.

   Write down the values of a and b.   2

(b) It is known that $\int_1^3 f(x)\,dx = 4$.

   Determine the value of $\int_6^8 h(x)\,dx$.   1

(c) Given $f'(1) = 6$, state the value of $h'(8)$.   1

[END OF QUESTION PAPER]

[BLANK PAGE]

DO NOT WRITE ON THIS PAGE

H National Qualifications 2017

X747/76/12

Mathematics Paper 2

FRIDAY, 5 MAY

10:30 AM – 12:00 NOON

Total marks — 70

Attempt ALL questions.

You may use a calculator.

Full credit will be given only to solutions which contain appropriate working.

State the units for your answer where appropriate.

Answers obtained by readings from scale drawings will not receive any credit.

Write your answers clearly in the spaces provided in the answer booklet. The size of the space provided for an answer should not be taken as an indication of how much to write. It is not necessary to use all the space.

Additional space for answers is provided at the end of the answer booklet. If you use this space **you must clearly identify the question number** you are attempting.

Use **blue** or **black** ink.

Before leaving the examination room you must give your answer booklet to the Invigilator; if you do not, you may lose all the marks for this paper.

FORMULAE LIST

Circle:

The equation $x^2 + y^2 + 2gx + 2fy + c = 0$ represents a circle centre $(-g, -f)$ and radius $\sqrt{g^2 + f^2 - c}$.

The equation $(x - a)^2 + (y - b)^2 = r^2$ represents a circle centre (a, b) and radius r.

Scalar Product: $\mathbf{a}.\mathbf{b} = |\mathbf{a}||\mathbf{b}| \cos \theta$, where θ is the angle between $\mathbf{a}$ and $\mathbf{b}$

or $\mathbf{a}.\mathbf{b} = a_1 b_1 + a_2 b_2 + a_3 b_3$ where $\mathbf{a} = \begin{pmatrix} a_1 \\ a_2 \\ a_3 \end{pmatrix}$ and $\mathbf{b} = \begin{pmatrix} b_1 \\ b_2 \\ b_3 \end{pmatrix}$.

Trigonometric formulae:
$$\sin (A \pm B) = \sin A \cos B \pm \cos A \sin B$$
$$\cos (A \pm B) = \cos A \cos B \mp \sin A \sin B$$
$$\sin 2A = 2 \sin A \cos A$$
$$\cos 2A = \cos^2 A - \sin^2 A$$
$$= 2 \cos^2 A - 1$$
$$= 1 - 2 \sin^2 A$$

Table of standard derivatives:

$f(x)$	$f'(x)$
$\sin ax$	$a \cos ax$
$\cos ax$	$-a \sin ax$

Table of standard integrals:

$f(x)$	$\int f(x)dx$
$\sin ax$	$-\dfrac{1}{a} \cos ax + c$
$\cos ax$	$\dfrac{1}{a} \sin ax + c$

MARKS

Attempt ALL questions

Total marks — 70

1. Triangle ABC is shown in the diagram below.

    The coordinates of B are (3,0) and the coordinates of C are (9,−2).

    The broken line is the perpendicular bisector of BC.

    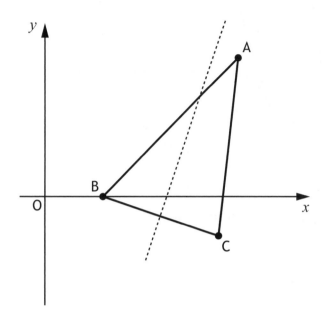

    (a) Find the equation of the perpendicular bisector of BC.   4

    (b) The line AB makes an angle of 45° with the positive direction of the x-axis.
         Find the equation of AB.   2

    (c) Find the coordinates of the point of intersection of AB and the perpendicular bisector of BC.   2

2. (a) Show that $(x-1)$ is a factor of $f(x)=2x^3-5x^2+x+2$.   2

    (b) Hence, or otherwise, solve $f(x)=0$.   3

[Turn over

MARKS

3. The line $y = 3x$ intersects the circle with equation $(x-2)^2 + (y-1)^2 = 25$.

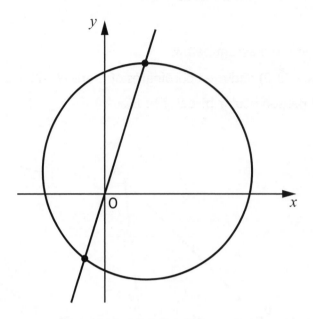

Find the coordinates of the points of intersection. 5

4. (a) Express $3x^2 + 24x + 50$ in the form $a(x+b)^2 + c$. 3

   (b) Given that $f(x) = x^3 + 12x^2 + 50x - 11$, find $f'(x)$. 2

   (c) Hence, or otherwise, explain why the curve with equation $y = f(x)$ is strictly increasing for all values of x. 2

MARKS

5. In the diagram, $\overrightarrow{PR} = 9\mathbf{i} + 5\mathbf{j} + 2\mathbf{k}$ and $\overrightarrow{RQ} = -12\mathbf{i} - 9\mathbf{j} + 3\mathbf{k}$.

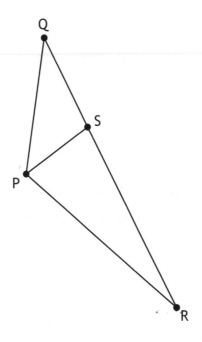

(a) Express $\overrightarrow{PQ}$ in terms of $\mathbf{i}$, $\mathbf{j}$ and $\mathbf{k}$.   2

The point S divides QR in the ratio 1:2.

(b) Show that $\overrightarrow{PS} = \mathbf{i} - \mathbf{j} + 4\mathbf{k}$.   2

(c) Hence, find the size of angle QPS.   5

6. Solve $5\sin x - 4 = 2\cos 2x$ for $0 \le x < 2\pi$.   5

7. (a) Find the x-coordinate of the stationary point on the curve
    with equation $y = 6x - 2\sqrt{x^3}$.   4

    (b) Hence, determine the greatest and least values of y in the interval $1 \le x \le 9$.   3

[Turn over

MARKS

8. Sequences may be generated by recurrence relations of the form
$u_{n+1} = k u_n - 20$, $u_0 = 5$ where $k \in \mathbb{R}$.

(a) Show that $u_2 = 5k^2 - 20k - 20$. 2

(b) Determine the range of values of k for which $u_2 < u_0$. 4

9. Two variables, x and y, are connected by the equation $y = kx^n$.

The graph of $\log_2 y$ against $\log_2 x$ is a straight line as shown.

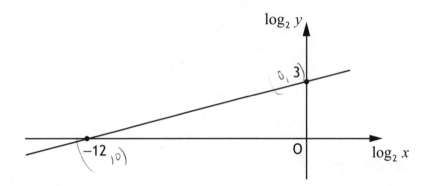

Find the values of k and n. 5

MARKS

10. (a) Show that the points A(−7, −2), B(2, 1) and C(17, 6) are collinear.   3

Three circles with centres A, B and C are drawn inside a circle with centre D as shown.

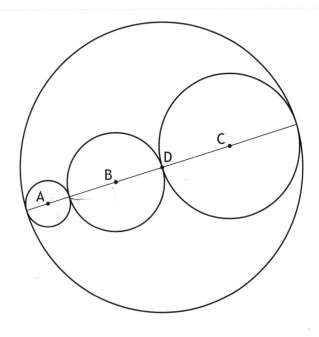

The circles with centres A, B and C have radii r_A, r_B and r_C respectively.

- $r_A = \sqrt{10}$

- $r_B = 2r_A$

- $r_C = r_A + r_B$

(b) Determine the equation of the circle with centre D.   4

11. (a) Show that $\dfrac{\sin 2x}{2\cos x} - \sin x \cos^2 x = \sin^3 x$, where $0 < x < \dfrac{\pi}{2}$.   3

(b) Hence, differentiate $\dfrac{\sin 2x}{2\cos x} - \sin x \cos^2 x$, where $0 < x < \dfrac{\pi}{2}$.   3

[END OF QUESTION PAPER]

[BLANK PAGE]

DO NOT WRITE ON THIS PAGE

HIGHER

2018

National Qualifications 2018

X747/76/11

Mathematics Paper 1 (Non-Calculator)

THURSDAY, 3 MAY

9:00 AM – 10:10 AM

Total marks — 60

Attempt ALL questions.

You may NOT use a calculator.

Full credit will be given only to solutions which contain appropriate working.

State the units for your answer where appropriate.

Answers obtained by readings from scale drawings will not receive any credit.

Write your answers clearly in the spaces provided in the answer booklet. The size of the space provided for an answer should not be taken as an indication of how much to write. It is not necessary to use all the space.

Additional space for answers is provided at the end of the answer booklet. If you use this space **you must clearly identify the question number** you are attempting.

Use **blue** or **black** ink.

Before leaving the examination room you must give your answer booklet to the Invigilator; if you do not, you may lose all the marks for this paper.

FORMULAE LIST

Circle:

The equation $x^2 + y^2 + 2gx + 2fy + c = 0$ represents a circle centre $(-g, -f)$ and radius $\sqrt{g^2 + f^2 - c}$.

The equation $(x - a)^2 + (y - b)^2 = r^2$ represents a circle centre (a, b) and radius r.

Scalar Product: $\mathbf{a}.\mathbf{b} = |\mathbf{a}||\mathbf{b}| \cos \theta$, where θ is the angle between $\mathbf{a}$ and $\mathbf{b}$

or $\mathbf{a}.\mathbf{b} = a_1 b_1 + a_2 b_2 + a_3 b_3$ where $\mathbf{a} = \begin{pmatrix} a_1 \\ a_2 \\ a_3 \end{pmatrix}$ and $\mathbf{b} = \begin{pmatrix} b_1 \\ b_2 \\ b_3 \end{pmatrix}$.

Trigonometric formulae:

$$\sin (A \pm B) = \sin A \cos B \pm \cos A \sin B$$
$$\cos (A \pm B) = \cos A \cos B \mp \sin A \sin B$$
$$\sin 2A = 2 \sin A \cos A$$
$$\cos 2A = \cos^2 A - \sin^2 A$$
$$= 2 \cos^2 A - 1$$
$$= 1 - 2 \sin^2 A$$

Table of standard derivatives:

$f(x)$	$f'(x)$
$\sin ax$	$a \cos ax$
$\cos ax$	$-a \sin ax$

Table of standard integrals:

$f(x)$	$\int f(x)dx$
$\sin ax$	$-\dfrac{1}{a} \cos ax + c$
$\cos ax$	$\dfrac{1}{a} \sin ax + c$

MARKS

Attempt ALL questions

Total marks — 60

1. PQR is a triangle with vertices P $(-2, 4)$, Q $(4, 0)$ and R $(3, 6)$.

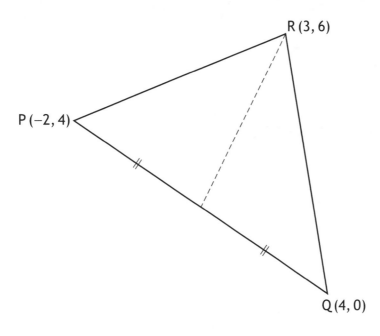

Find the equation of the median through R.   3

2. A function $g(x)$ is defined on $\mathbb{R}$, the set of real numbers, by

$$g(x) = \frac{1}{5}x - 4.$$

Find the inverse function, $g^{-1}(x)$.   3

3. Given $h(x) = 3\cos 2x$, find the value of $h'\left(\frac{\pi}{6}\right)$.   3

[Turn over

MARKS

4. The point K (8, −5) lies on the circle with equation $x^2 + y^2 - 12x - 6y - 23 = 0$.

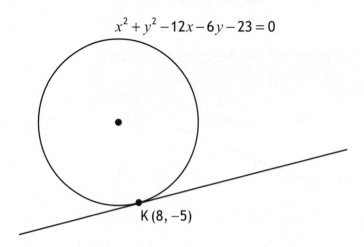

$$x^2 + y^2 - 12x - 6y - 23 = 0$$

K (8, −5)

Find the equation of the tangent to the circle at K.   4

5. A(−3, 4, −7), B (5, t, 5) and C (7, 9, 8) are collinear.

    (a) State the ratio in which B divides AC.   1

    (b) State the value of t.   1

6. Find the value of $\log_5 250 - \dfrac{1}{3}\log_5 8$.   3

MARKS

7. The curve with equation $y = x^3 - 3x^2 + 2x + 5$ is shown on the diagram.

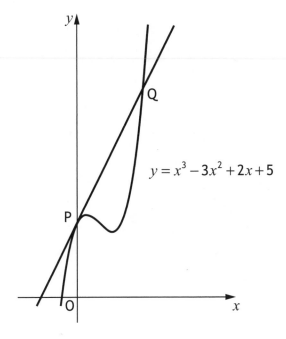

$y = x^3 - 3x^2 + 2x + 5$

(a) Write down the coordinates of P, the point where the curve crosses the y-axis. **1**

(b) Determine the equation of the tangent to the curve at P. **3**

(c) Find the coordinates of Q, the point where this tangent meets the curve again. **4**

8. A line has equation $y - \sqrt{3}\,x + 5 = 0$.

Determine the angle this line makes with the positive direction of the x-axis. **2**

[**Turn over**

MARKS

9. The diagram shows a triangular prism ABC,DEF.

    $\overrightarrow{AB} = \mathbf{t}$, $\overrightarrow{AC} = \mathbf{u}$ and $\overrightarrow{AD} = \mathbf{v}$.

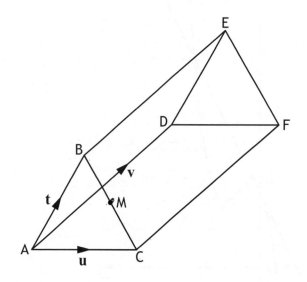

    (a) Express $\overrightarrow{BC}$ in terms of $\mathbf{u}$ and $\mathbf{t}$.   1

    M is the midpoint of BC.

    (b) Express $\overrightarrow{MD}$ in terms of $\mathbf{t}$, $\mathbf{u}$ and $\mathbf{v}$.   2

10. Given that

    • $\dfrac{dy}{dx} = 6x^2 - 3x + 4$, and

    • $y = 14$ when $x = 2$,

    express y in terms of x.   4

MARKS

11. The diagram shows the curve with equation $y = \log_3 x$.

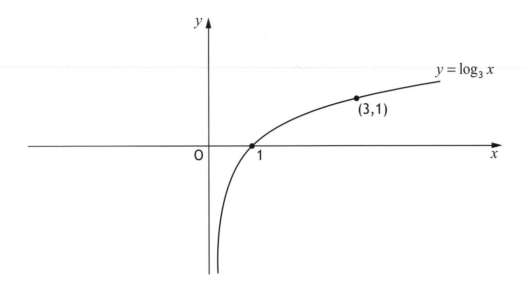

(a) On the diagram in your answer booklet, sketch the curve with equation $y = 1 - \log_3 x$.   2

(b) Determine the exact value of the x-coordinate of the point of intersection of the two curves.   3

12. Vectors **a** and **b** are such that $\mathbf{a} = 4\mathbf{i} - 2\mathbf{j} + 2\mathbf{k}$ and $\mathbf{b} = -2\mathbf{i} + \mathbf{j} + p\mathbf{k}$.

(a) Express $2\mathbf{a} + \mathbf{b}$ in component form.   1

(b) Hence find the values of p for which $|2\mathbf{a} + \mathbf{b}| = 7$.   3

[Turn over for next question

MARKS

13. The right-angled triangle in the diagram is such that $\sin x = \dfrac{2}{\sqrt{11}}$ and $0 < x < \dfrac{\pi}{4}$.

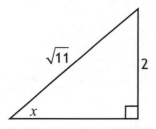

(a) Find the exact value of:

    (i) $\sin 2x$ **3**

    (ii) $\cos 2x$.   **1**

(b) By expressing $\sin 3x$ as $\sin(2x + x)$, find the exact value of $\sin 3x$.   **3**

14. Evaluate $\displaystyle\int_{-4}^{9} \dfrac{1}{\sqrt[3]{(2x+9)^2}}\,dx$.   **5**

15. A cubic function, f, is defined on the set of real numbers.

- $(x+4)$ is a factor of $f(x)$

- $x = 2$ is a repeated root of $f(x)$

- $f'(-2) = 0$

- $f'(x) > 0$ where the graph with equation $y = f(x)$ crosses the y-axis

Sketch a possible graph of $y = f(x)$ on the diagram in your answer booklet.   **4**

[END OF QUESTION PAPER]

National Qualifications 2018

X747/76/12

Mathematics Paper 2

THURSDAY, 3 MAY

10:30 AM – 12:00 NOON

Total marks — 70

Attempt ALL questions.

You may use a calculator.

Full credit will be given only to solutions which contain appropriate working.

State the units for your answer where appropriate.

Answers obtained by readings from scale drawings will not receive any credit.

Write your answers clearly in the spaces provided in the answer booklet. The size of the space provided for an answer should not be taken as an indication of how much to write. It is not necessary to use all the space.

Additional space for answers is provided at the end of the answer booklet. If you use this space **you must clearly identify the question number** you are attempting.

Use **blue** or **black** ink.

Before leaving the examination room you must give your answer booklet to the Invigilator; if you do not, you may lose all the marks for this paper.

FORMULAE LIST

Circle:

The equation $x^2 + y^2 + 2gx + 2fy + c = 0$ represents a circle centre $(-g, -f)$ and radius $\sqrt{g^2 + f^2 - c}$.

The equation $(x - a)^2 + (y - b)^2 = r^2$ represents a circle centre (a, b) and radius r.

Scalar Product: $\mathbf{a}.\mathbf{b} = |\mathbf{a}||\mathbf{b}| \cos \theta$, where θ is the angle between $\mathbf{a}$ and $\mathbf{b}$

or $\mathbf{a}.\mathbf{b} = a_1 b_1 + a_2 b_2 + a_3 b_3$ where $\mathbf{a} = \begin{pmatrix} a_1 \\ a_2 \\ a_3 \end{pmatrix}$ and $\mathbf{b} = \begin{pmatrix} b_1 \\ b_2 \\ b_3 \end{pmatrix}$.

Trigonometric formulae:

$$\sin (A \pm B) = \sin A \cos B \pm \cos A \sin B$$
$$\cos (A \pm B) = \cos A \cos B \mp \sin A \sin B$$
$$\sin 2A = 2 \sin A \cos A$$
$$\cos 2A = \cos^2 A - \sin^2 A$$
$$= 2 \cos^2 A - 1$$
$$= 1 - 2 \sin^2 A$$

Table of standard derivatives:

$f(x)$	$f'(x)$
$\sin ax$	$a \cos ax$
$\cos ax$	$-a \sin ax$

Table of standard integrals:

$f(x)$	$\int f(x)dx$
$\sin ax$	$-\dfrac{1}{a} \cos ax + c$
$\cos ax$	$\dfrac{1}{a} \sin ax + c$

MARKS

Attempt ALL questions

Total marks — 70

1. The diagram shows the curve with equation $y = 3 + 2x - x^2$.

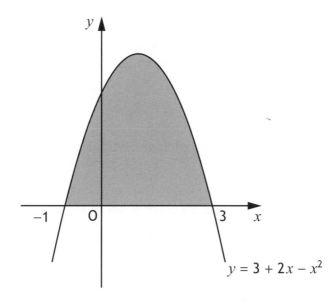

    Calculate the shaded area.   4

2. Vectors **u** and **v** are defined by $\mathbf{u} = \begin{pmatrix} -1 \\ 4 \\ -3 \end{pmatrix}$ and $\mathbf{v} = \begin{pmatrix} -7 \\ 8 \\ 5 \end{pmatrix}$.

    (a) Find **u.v**.   1

    (b) Calculate the acute angle between **u** and **v**.   4

3. A function, f, is defined on the set of real numbers by $f(x) = x^3 - 7x - 6$.

    Determine whether f is increasing or decreasing when $x = 2$.   3

4. Express $-3x^2 - 6x + 7$ in the form $a(x+b)^2 + c$.   3

[Turn over

MARKS

5.   PQR is a triangle with P(3,4) and Q(9,−2).

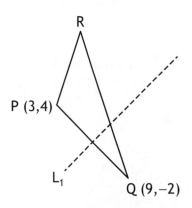

   (a) Find the equation of L_1, the perpendicular bisector of PQ.   3

The equation of L_2, the perpendicular bisector of PR is $3y + x = 25$.

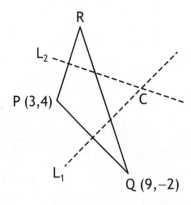

   (b) Calculate the coordinates of C, the point of intersection of L_1 and L_2.   2

C is the centre of the circle which passes through the vertices of triangle PQR.

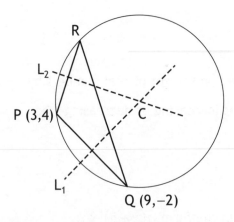

   (c) Determine the equation of this circle.   2

MARKS

6. Functions, f and g, are given by $f(x) = 3 + \cos x$ and $g(x) = 2x$, $x \in \mathbb{R}$.

   (a) Find expressions for

       (i) $f(g(x))$ and   2

       (ii) $g(f(x))$.   1

   (b) Determine the value(s) of x for which $f(g(x)) = g(f(x))$ where $0 \leq x < 2\pi$.   6

7. (a)   (i) Show that $(x - 2)$ is a factor of $2x^3 - 3x^2 - 3x + 2$.   2

      (ii) Hence, factorise $2x^3 - 3x^2 - 3x + 2$ fully.   2

The fifth term, u_5, of a sequence is $u_5 = 2a - 3$.

The terms of the sequence satisfy the recurrence relation $u_{n+1} = au_n - 1$.

   (b) Show that $u_7 = 2a^3 - 3a^2 - a - 1$.   1

For this sequence, it is known that

- $u_7 = u_5$
- a limit exists.

   (c)   (i) Determine the value of a.   3

      (ii) Calculate the limit.   1

[Turn over

MARKS

8. (a) Express $2\cos x° - \sin x°$ in the form $k\cos(x-a)°$, $k>0$, $0<a<360$.   4

   (b) Hence, or otherwise, find

       (i) the minimum value of $6\cos x° - 3\sin x°$ and   1

       (ii) the value of x for which it occurs where $0 \le x < 360$.   2

9. A sector with a particular fixed area has radius x cm.

   The perimeter, P cm, of the sector is given by

   $$P = 2x + \frac{128}{x}.$$

   Find the minimum value of P.   6

10. The equation $x^2 + (m-3)x + m = 0$ has two real and distinct roots.

    Determine the range of values for m.   4

11. A supermarket has been investigating how long customers have to wait at the checkout.

    During any half hour period, the percentage, $P\%$, of customers who wait for less than t minutes, can be modelled by

    $$P = 100\left(1 - e^{kt}\right), \text{ where } k \text{ is a constant.}$$

    (a) If 50% of customers wait for less than 3 minutes, determine the value of k.   4

    (b) Calculate the percentage of customers who wait for 5 minutes or longer.   2

12. **Circle C_1 has equation $(x-13)^2 + (y+4)^2 = 100$.**

Circle C_2 has equation $x^2 + y^2 + 14x - 22y + c = 0$.

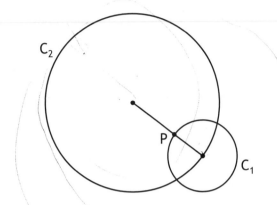

(a) (i) Write down the coordinates of the centre of C_1. 1

(ii) The centre of C_1 lies on the circumference of C_2.

Show that $c = -455$. 1

The line joining the centres of the circles intersects C_1 at P.

(b) (i) Determine the ratio in which P divides the line joining the centres of the circles. 2

(ii) Hence, or otherwise, determine the coordinates of P. 2

P is the centre of a third circle, C_3.

C_2 touches C_3 internally.

(c) Determine the equation of C_3. 1

[END OF QUESTION PAPER]

[BLANK PAGE]

DO NOT WRITE ON THIS PAGE

National
Qualifications
SPECIMEN ONLY

S847/76/11

**Mathematics
Paper 1
(Non-Calculator)**

Date — Not applicable

Duration — 1 hour 30 minutes

Total marks — 70

Attempt ALL questions.

You may NOT use a calculator.

To earn full marks you must show your working in your answers.

State the units for your answer where appropriate.

You will not earn marks for answers obtained by readings from scale drawings.

Write your answers clearly in the spaces provided in the answer booklet. The size of the space provided for an answer is not an indication of how much to write. You do not need to use all the space.

Additional space for answers is provided at the end of the answer booklet. If you use this space you must clearly identify the question number you are attempting.

Use **blue** or **black** ink.

Before leaving the examination room you must give your answer booklet to the Invigilator; if you do not, you may lose all the marks for this paper.

FORMULAE LIST

Circle:

The equation $x^2 + y^2 + 2gx + 2fy + c = 0$ represents a circle centre $(-g, -f)$ and radius $\sqrt{g^2 + f^2 - c}$.

The equation $(x - a)^2 + (y - b)^2 = r^2$ represents a circle centre (a, b) and radius r.

Scalar Product: $\mathbf{a}.\mathbf{b} = |\mathbf{a}||\mathbf{b}| \cos \theta$, where θ is the angle between $\mathbf{a}$ and $\mathbf{b}$

or $\mathbf{a}.\mathbf{b} = a_1 b_1 + a_2 b_2 + a_3 b_3$ where $\mathbf{a} = \begin{pmatrix} a_1 \\ a_2 \\ a_3 \end{pmatrix}$ and $\mathbf{b} = \begin{pmatrix} b_1 \\ b_2 \\ b_3 \end{pmatrix}$.

Trigonometric formulae:

$$\sin (A \pm B) = \sin A \cos B \pm \cos A \sin B$$
$$\cos (A \pm B) = \cos A \cos B \mp \sin A \sin B$$
$$\sin 2A = 2 \sin A \cos A$$
$$\cos 2A = \cos^2 A - \sin^2 A$$
$$= 2 \cos^2 A - 1$$
$$= 1 - 2 \sin^2 A$$

Table of standard derivatives:

$f(x)$	$f'(x)$
$\sin ax$	$a \cos ax$
$\cos ax$	$-a \sin ax$

Table of standard integrals:

$f(x)$	$\int f(x)dx$
$\sin ax$	$-\dfrac{1}{a} \cos ax + c$
$\cos ax$	$\dfrac{1}{a} \sin ax + c$

MARKS

Attempt ALL questions

Total marks — 70

1. A curve has equation $y = x^2 - 4x + 7$.

    Find the equation of the tangent to this curve at the point where $x = 5$.   4

2. A and B are the points $(-7, 3)$ and $(1, 5)$.

    AB is a diameter of a circle.

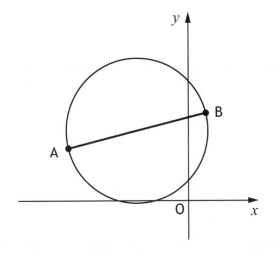

    Find the equation of this circle.   3

3. Line l_1 has equation $\sqrt{3}y - x = 0$.

    (a) Line l_2 is perpendicular to l_1. Find the gradient of l_2.   2

    (b) Calculate the angle l_2 makes with the positive direction of the x-axis.   2

MARKS

4. Evaluate $\int_{1}^{2} \frac{1}{6} x^{-2} \, dx.$

3

5. The points $A(0, 9, 7)$, $B(5, -1, 2)$, $C(4, 1, 3)$ and $D(x, -2, 2)$ are such that $\overrightarrow{AB}$ is perpendicular to $\overrightarrow{CD}$.

Determine the value of x.

4

6. Determine the range of values of p such that the equation $x^2 + (p+1)x + 9 = 0$ has no real roots.

4

7. Show that the line with equation $y = 3x - 5$ is a tangent to the circle with equation $x^2 + y^2 + 2x - 4y - 5 = 0$ and find the coordinates of the point of contact.

5

MARKS

8. For the polynomial, $x^3 - 4x^2 + ax + b$

   - $x - 1$ is a factor
   - -12 is the remainder when it is divided by $x - 2$

   (a) Determine the values of a and b.   5

   (b) Hence solve $x^3 - 4x^2 + ax + b = 0$.   3

9. A sequence is generated by the recurrence relation $u_{n+1} = m u_n + 6$ where m is a constant.

   (a) Given $u_1 = 28$ and $u_2 = 13$, find the value of m.   2

   (b) (i) Explain why this sequence approaches a limit as $n \to \infty$.   1

   (ii) Calculate this limit.   2

10. (a) Evaluate $\log_5 25$.   1

    (b) Hence solve $\log_4 x + \log_4 (x - 6) = \log_5 25$, where $x > 6$.   5

11. Find the rate of change of the function $f(x) = 4\sin^3 x$ when $x = \dfrac{5\pi}{6}$.   3

MARKS

12. Triangle ABD is right-angled at B with angles BAC $= p$ and BAD $= q$ and lengths as shown in the diagram below.

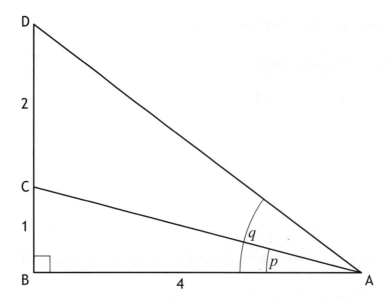

Show that the exact value of $\cos(q - p)$ is $\dfrac{19\sqrt{17}}{85}$.

5

13. The curve $y = f(x)$ is such that $\dfrac{dy}{dx} = 4x - 6x^2$. The curve passes through the point $(-1, 9)$. Express y in terms of x.

4

14. (a) Solve $\cos 2x° - 3\cos x° + 2 = 0$ for $0 \le x < 360$.

5

(b) Hence solve $\cos 4x° - 3\cos 2x° + 2 = 0$ for $0 \le x < 360$.

2

MARKS

15. Functions f and g are defined on suitable domains by $f(x) = x^3 - 1$ and $g(x) = 3x + 1$.

   (a) Find an expression for $k(x)$, where $k(x) = g(f(x))$.   2

   (b) If $h(k(x)) = x$, find an expression for $h(x)$.   3

[END OF SPECIMEN QUESTION PAPER]

[BLANK PAGE]

DO NOT WRITE ON THIS PAGE

National
Qualifications
SPECIMEN ONLY

S847/76/12

**Mathematics
Paper 2**

Date — Not applicable

Duration — 1 hour 45 minutes

Total marks — 80

Attempt ALL questions.

You may use a calculator.

To earn full marks you must show your working in your answers.

State the units for your answer where appropriate.

You will not earn marks for answers obtained by readings from scale drawings.

Write your answers clearly in the spaces provided in the answer booklet. The size of the space provided for an answer is not an indication of how much to write. You do not need to use all the space.

Additional space for answers is provided at the end of the answer booklet. If you use this space you must clearly identify the question number you are attempting.

Use **blue** or **black** ink.

Before leaving the examination room you must give your answer booklet to the Invigilator; if you do not, you may lose all the marks for this paper.

FORMULAE LIST

Circle:

The equation $x^2 + y^2 + 2gx + 2fy + c = 0$ represents a circle centre $(-g, -f)$ and radius $\sqrt{g^2 + f^2 - c}$.

The equation $(x - a)^2 + (y - b)^2 = r^2$ represents a circle centre (a, b) and radius r.

Scalar Product: $\mathbf{a.b} = |\mathbf{a}||\mathbf{b}| \cos \theta$, where θ is the angle between $\mathbf{a}$ and $\mathbf{b}$

or $\mathbf{a.b} = a_1 b_1 + a_2 b_2 + a_3 b_3$ where $\mathbf{a} = \begin{pmatrix} a_1 \\ a_2 \\ a_3 \end{pmatrix}$ and $\mathbf{b} = \begin{pmatrix} b_1 \\ b_2 \\ b_3 \end{pmatrix}$.

Trigonometric formulae:

$$\sin (A \pm B) = \sin A \cos B \pm \cos A \sin B$$
$$\cos (A \pm B) = \cos A \cos B \mp \sin A \sin B$$
$$\sin 2A = 2 \sin A \cos A$$
$$\cos 2A = \cos^2 A - \sin^2 A$$
$$= 2 \cos^2 A - 1$$
$$= 1 - 2 \sin^2 A$$

Table of standard derivatives:

$f(x)$	$f'(x)$
$\sin ax$	$a \cos ax$
$\cos ax$	$-a \sin ax$

Table of standard integrals:

$f(x)$	$\int f(x)dx$
$\sin ax$	$-\dfrac{1}{a} \cos ax + c$
$\cos ax$	$\dfrac{1}{a} \sin ax + c$

MARKS

Attempt ALL questions

Total marks — 80

1. The vertices of triangle ABC are A(−5, 7), B(−1, −5) and C(13, 3) as shown in the diagram.

   The broken line represents the altitude from C.

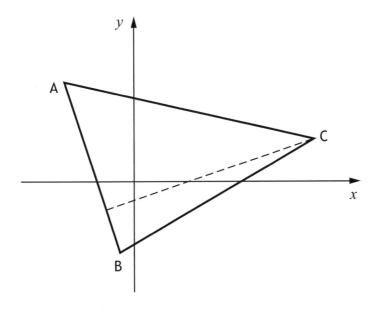

   (a) Find the equation of the altitude from C.   3

   (b) Find the equation of the median from B.   3

   (c) Find the coordinates of the point of intersection of the altitude from C and the median from B.   2

2. Find $\int \dfrac{4x^3 + 1}{x^2}\,dx,\ x \neq 0.$ 4

MARKS

3. The diagram shows the curve with equation $y = f(x)$, where
    $f(x) = kx(x+a)(x+b)$.

    The curve passes through $(-1, 0)$, $(0, 0)$, $(1, 2)$ and $(2, 0)$.

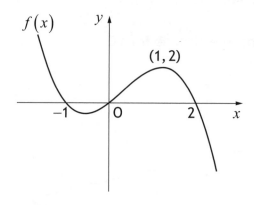

    Find the values of a, b and k.   3

4. D,OABC is a square-based pyramid as shown.

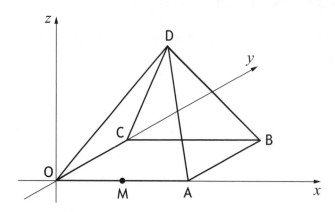

    •   O is the origin and OA $= 4$ units.

    •   M is the mid-point of OA.

    • $\overrightarrow{OD} = 2\mathbf{i} + 2\mathbf{j} + 6\mathbf{k}$

    (a) Express $\overrightarrow{DB}$ and $\overrightarrow{DM}$ in component form.   3

    (b) Find the size of angle BDM.   5

MARKS

5. The line with equation $y = 2x + 3$ is a tangent to the curve with equation $y = x^3 + 3x^2 + 2x + 3$ at A(0, 3), as shown.

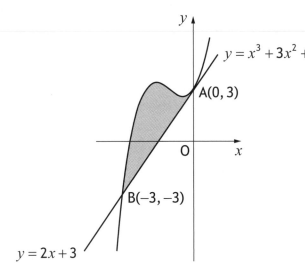

The line meets the curve again at B(−3, −3).

Find the area enclosed by the line and the curve.   5

6. (a) Express $3x^2 + 24x + 50$ in the form $a(x+b)^2 + c$.   3

   (b) Given that $f(x) = x^3 + 12x^2 + 50x - 11$, find $f'(x)$.   2

   (c) Hence, or otherwise, explain why the curve with equation $y = f(x)$ is strictly
       increasing for all values of x.   2

MARKS

7. The diagram below shows the graph of a quartic $y = h(x)$, with stationary points at $(0, 5)$ and $(2, 2)$.

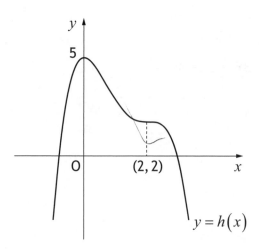

On separate diagrams sketch the graphs of:

(a) $y = 2 - h(x)$.   2

(b) $y = h'(x)$.   3

8. (a) Express $5\cos x - 2\sin x$ in the form $k\cos(x+a)$, where $k > 0$ and $0 < a < 2\pi$.   4

    (b) The diagram shows a sketch of part of the graph of $y = 10 + 5\cos x - 2\sin x$ and the line with equation $y = 12$.

    The line cuts the curve at the points P and Q.

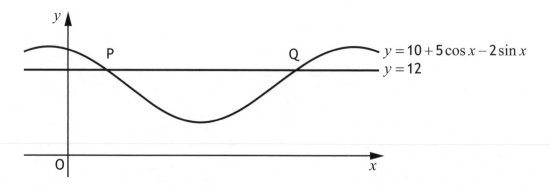

    Find the x-coordinates of P and Q.   4

MARKS

9. A design for a new grain container is in the shape of a cylinder with a hemispherical roof and a flat circular base. The radius of the cylinder is r metres, and the height is h metres.

The volume of the **cylindrical** part of the container needs to be 100 cubic metres.

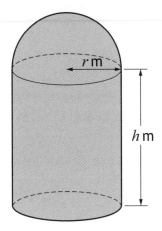

(a) Given that the curved surface area of a hemisphere of radius r is $2\pi r^2$ show that the surface area of metal needed to build the grain container is given by:

$$A = \frac{200}{r} + 3\pi r^2 \text{ square metres}$$

3

(b) Determine the value of r which minimises the amount of metal needed to build the container.

6

10. Given that

$$\int_{\frac{\pi}{8}}^{a} \sin\left(4x - \frac{\pi}{2}\right) dx = \frac{1}{2}, \quad 0 \le a < \frac{\pi}{2},$$

calculate the value of a.

6

MARKS

11. Show that $\dfrac{\sin 2x}{2\cos x} - \sin x\cos^2 x = \sin^3 x$, where $0 < x < \dfrac{\pi}{2}$.

3

12. (a) Show that the points A$(-7, -2)$, B$(2, 1)$ and C$(17, 6)$ are collinear.

3

Three circles with centres A, B and C are drawn inside a circle with centre D as shown.

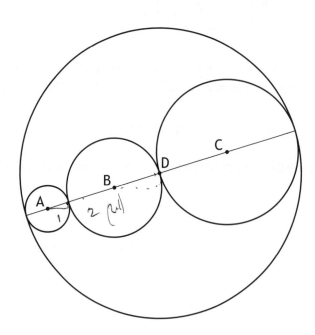

The circles with centres A, B and C have radii r_A, r_B and r_C respectively.

- $r_A = \sqrt{10}$

- $r_B = 2r_A$

- $r_C = r_A + r_B$

(b) Determine the equation of the circle with centre D.

4

MARKS

13. The concentration of a pesticide in soil can be modelled by the equation

$$P_t = P_0 e^{-kt}$$

where:

- P_0 is the initial concentration;
- P_t is the concentration at time t;
- t is the time, in days, after the application of the pesticide.

(a) It takes 25 days for the concentration of the pesticide to be reduced to one half of its initial concentration.

Calculate the value of k.   4

(b) Eighty days after the initial application, what is the percentage decrease in concentration of the pesticide?   3

[END OF SPECIMEN QUESTION PAPER]

[BLANK PAGE]

DO NOT WRITE ON THIS PAGE

HIGHER

Answers

ANSWERS FOR

SQA HIGHER MATHEMATICS 2018

HIGHER MATHEMATICS
2017

Paper 1 (Non-Calculator)

Question			Generic Scheme	Illustrative Scheme	Max mark
1.	(a)		•^{1} evaluate expression	•^{1} 10	1
	(b)		•^{2} interpret notation	•^{2} $g(5x)$	2
			•^{3} state expression for $g(f(x))$	•^{3} $2\cos 5x$	
2.			•^{1} state coordinates of centre	•^{1} $(4,3)$	4
			•^{2} find gradient of radius	•^{2} $\dfrac{1}{3}$	
			•^{3} state perpendicular gradient	•^{3} -3	
			•^{4} determine equation of tangent	•^{4} $y=-3x-5$	
3.			•^{1} start to differentiate	•^{1} $12(4x-1)^{11}$	2
			•^{2} complete differentiation	•^{2} ....×4	
4.			**Method 1**	**Method 1**	3
			•^{1} use the discriminant	•^{1} $4^2-4\times 1\times(k-5)$	
			•^{2} apply condition and simplify	•^{2} $36-4k=0$ or $36=4k$	
			•^{3} determine the value of k	•^{3} $k=9$	
			Method 2	**Method 2**	
			•^{1} communicate and express in factorised form	•^{1} equal roots $\Rightarrow x^2+4x+(k-5)=(x+2)^2$	
			•^{2} expand and compare	•^{2} x^2+4x+4 leading to $k-5=4$	
			•^{3} determine the value of k	•^{3} $k=9$	
5.	(a)		•^{1} evaluate scalar product	•^{1} 1	1
	(b)		•^{2} calculate $|\mathbf{u}|$	•^{2} $\sqrt{27}$	3
			•^{3} use scalar product	•^{3} $\sqrt{27}\times\sqrt{3}\times\cos\dfrac{\pi}{3}$	
			•^{4} evaluate $\mathbf{u}\cdot\mathbf{w}$	•^{4} $\dfrac{9}{2}$ or 4·5	

Question			Generic Scheme	Illustrative Scheme	Max mark
6.			**Method 1**	**Method 1**	3
			$\bullet^1$ equate composite function to x	$\bullet^1$ $h\left(h^{-1}(x)\right)=x$	
			$\bullet^2$ write $h\left(h^{-1}(x)\right)$ in terms of $h^{-1}(x)$	$\bullet^2$ $\left(h^{-1}(x)\right)^3+7=x$	
			$\bullet^3$ state inverse function	$\bullet^3$ $h^{-1}(x)=\sqrt[3]{x-7}$ or $h^{-1}(x)=(x-7)^{\frac{1}{3}}$	
			Method 2	**Method 2**	
			$\bullet^1$ write as $y=x^3+7$ and start to rearrange	$\bullet^1$ $y-7=x^3$	
			$\bullet^2$ complete rearrangement	$\bullet^2$ $x=\sqrt[3]{y-7}$	
			$\bullet^3$ state inverse function	$\bullet^3$ $h^{-1}(x)=\sqrt[3]{x-7}$ or $h^{-1}(x)=(x-7)^{\frac{1}{3}}$	
			Method 3	**Method 3**	
			$\bullet^1$ interchange variables	$\bullet^1$ $x=y^3+7$	
			$\bullet^2$ complete rearrangement	$\bullet^2$ $y=\sqrt[3]{x-7}$	
			$\bullet^3$ state inverse function	$\bullet^3$ $h^{-1}(x)=\sqrt[3]{x-7}$ or $h^{-1}(x)=(x-7)^{\frac{1}{3}}$	
7.			$\bullet^1$ find midpoint of AB	$\bullet^1$ $(2,7)$	3
			$\bullet^2$ demonstrate the line is vertical	$\bullet^2$ m_{median} undefined	
			$\bullet^3$ state equation	$\bullet^3$ $x=2$	
8.			$\bullet^1$ write in differentiable form	$\bullet^1$ $\frac{1}{2}t^{-1}$	3
			$\bullet^2$ differentiate	$\bullet^2$ $-\frac{1}{2}t^{-2}$	
			$\bullet^3$ evaluate derivative	$\bullet^3$ $-\frac{1}{50}$	
9.	(a)		$\bullet^1$ interpret information	$\bullet^1$ $13=28m+6$ **stated explicitly or in a rearranged form**	2
			$\bullet^2$ state the value of m	$\bullet^2$ $m=\frac{1}{4}$ or $m=0\!\cdot\!25$	
	(b)	(i)	$\bullet^3$ communicate condition for limit to exist	$\bullet^3$ a limit exists as the recurrence relation is linear and $-1<\frac{1}{4}<1$	1
		(ii)	$\bullet^4$ know how to calculate limit	$\bullet^4$ $\dfrac{6}{1-\frac{1}{4}}$ or $L=\frac{1}{4}L+6$	2
			$\bullet^5$ calculate limit	$\bullet^5$ 8	

Question			Generic Scheme	Illustrative Scheme	Max mark
10.	(a)		**Method 1**	**Method 1**	5
			•¹ know to integrate between appropriate limits	•¹ $\int_0^2 \dots dx$	
			•² use *"upper – lower"*	•² $\int_0^2 \left((x^3 - 4x^2 + 3x + 1) - (x^2 - 3x + 1) \right)$	
			•³ integrate	•³ $\dfrac{x^4}{4} - \dfrac{5x^3}{3} + 3x^2$	
			•⁴ substitute limits	•⁴ $\left(\dfrac{2^4}{4} - \dfrac{5 \times 2^3}{3} + 3 \times 2^2 \right) - (0)$	
			•⁵ evaluate area	•⁵ $\dfrac{8}{3}$	
			Method 2	**Method 2**	
			•¹ know to integrate between appropriate limits for both integrals	•¹ $\int_0^2 \dots dx$ and $\int_0^2 \dots dx$	
			•² integrate both functions	•² $\dfrac{x^4}{4} - \dfrac{4x^3}{3} + \dfrac{3x^2}{2} + x$ and $\dfrac{x^3}{3} - \dfrac{3x^2}{2} + x$	
			•³ substitute limits into both functions	•³ $\left(\dfrac{2^4}{4} - \dfrac{4(2^3)}{3} + \dfrac{3(2^2)}{2} + 2 \right) - 0$ and $\left(\dfrac{2^3}{3} - \dfrac{3(2^2)}{2} + 2 \right) - 0$	
			•⁴ evaluation of both functions	•⁴ $\dfrac{4}{3}$ and $\dfrac{-4}{3}$	
			•⁵ evidence of subtracting areas	•⁵ $\dfrac{4}{3} - \dfrac{-4}{3} = \dfrac{8}{3}$	

Question			Generic Scheme	Illustrative Scheme	Max mark
10.	(b)		**Method 1**	**Method 1**	4
			$\bullet^6$ use "*line – quadratic*"	$\bullet^6 \int \left((1-x) - (x^2 - 3x + 1) \right) dx$	
			$\bullet^7$ integrate	$\bullet^7 \; -\dfrac{x^3}{3} + x^2$	
			$\bullet^8$ substitute limits and evaluate integral	$\bullet^8 \; \left(-\dfrac{2^3}{3} + 2^2 \right) - (0) = \dfrac{4}{3}$	
			$\bullet^9$ state fraction	$\bullet^9 \; \dfrac{1}{2}$	
			Method 2	**Method 2**	
			$\bullet^6$ use "*cubic - line*"	$\bullet^6 \int \left((x^3 - 4x^2 + 3x + 1) - (1-x) \right) dx$	
			$\bullet^7$ integrate	$\bullet^7 \; \dfrac{x^4}{4} - \dfrac{4x^3}{3} + 2x^2$	
			$\bullet^8$ substitute limits and evaluate integral	$\bullet^8 \; \left(\dfrac{2^4}{4} - 4 \times \dfrac{2^3}{3} + 2 \times 2^2 \right) - (0) = \dfrac{4}{3}$	
			$\bullet^9$ state fraction	$\bullet^9 \; \dfrac{1}{2}$	
			Method 3	**Method 3**	
			$\bullet^6$ integrate line	$\bullet^6 \int (1-x)\, dx = \left[x - \dfrac{x^2}{2} \right]_0^2$	
			$\bullet^7$ substitute limits and evaluate integral	$\bullet^7 \; \left(2 - \dfrac{2^2}{2} \right) - (0) = 0$	
			$\bullet^8$ evidence of subtracting integrals	$\bullet^8 \; 0 - \left(-\dfrac{4}{3} \right) = \dfrac{4}{3} \; \text{or} \; \dfrac{4}{3} - 0$	
			$\bullet^9$ state fraction	$\bullet^9 \; \dfrac{1}{2}$	

Question	Generic Scheme	Illustrative Scheme	Max mark
11.	**Method 1**	**Method 1**	**3**
	$\bullet^1$ determine the gradient of given line or of AB	$\bullet^1$ $\dfrac{2}{3}$ or $\dfrac{a-2}{12}$	
	$\bullet^2$ determine the other gradient	$\bullet^2$ $\dfrac{a-2}{12}$ or $\dfrac{2}{3}$	
	$\bullet^3$ find a	$\bullet^3$ 10	
	Method 2	**Method 2**	
	$\bullet^1$ determine the gradient of given line	$\bullet^1$ $\dfrac{2}{3}$ stated or implied by $\bullet^2$	
	$\bullet^2$ equation of line and substitute	$\bullet^2$ $y-2=\dfrac{2}{3}(x+7)$ $a-2=\dfrac{2}{3}(5+7)$	
	$\bullet^3$ solve for a	$\bullet^3$ 10	
12.	$\bullet^1$ use laws of logs	$\bullet^1$ $\log_a 9$	**3**
	$\bullet^2$ write in exponential form	$\bullet^2$ $a^{\frac{1}{2}}=9$	
	$\bullet^3$ solve for a	$\bullet^3$ 81	
13.	$\bullet^1$ write in integrable form	$\bullet^1$ $(5-4x)^{-\frac{1}{2}}$	**4**
	$\bullet^2$ start to integrate	$\bullet^2$ $\dfrac{(5-4x)^{\frac{1}{2}}}{\frac{1}{2}}$ ...	
	$\bullet^3$ process coefficient of x	$\bullet^3$ ...$\times\dfrac{1}{(-4)}$	
	$\bullet^4$ complete integration and simplify	$\bullet^4$ $-\dfrac{1}{2}(5-4x)^{\frac{1}{2}}+c$	

Question			Generic Scheme	Illustrative Scheme	Max mark
14.	(a)		•^{1} use compound angle formula	•^{1} $k\sin x°\cos a° - k\cos x°\sin a°$ **stated explicitly**	4
			•^{2} compare coefficients	•^{2} $k\cos a° = \sqrt{3},\ k\sin a° = 1$ **stated explicitly**	
			•^{3} process for k	•^{3} $k = 2$	
			•^{4} process for a and express in required form	•^{4} $2\sin(x-30)°$	
	(b)		•^{5} roots identifiable from graph	•^{5} 30 and 210	3
			•^{6} coordinates of both turning points identifiable from graph	•^{6} $(120, 2)$ and $(300, -2)$	
			•^{7} y-intercept and value of y at $x = 360$ identifiable from graph	•^{7} -1	
15.	(a)		•^{1} state value of a	•^{1} -5	2
			•^{2} state value of b	•^{2} 3	
	(b)		•^{3} state value of integral	•^{3} 10	1
	(c)		•^{4} state value of derivative	•^{4} -6	1

Paper 2 (Calculator)

Question			Generic Scheme	Illustrative Scheme	Max mark
1.	(a)		$\bullet^1$ find mid-point of BC	$\bullet^1$ $(6,-1)$	4
			$\bullet^2$ calculate gradient of BC	$\bullet^2$ $-\dfrac{2}{6}$	
			$\bullet^3$ use property of perpendicular lines	$\bullet^3$ 3	
			$\bullet^4$ determine equation of line in a simplified form	$\bullet^4$ $y=3x-19$	
	(b)		$\bullet^5$ use $m=\tan\theta$	$\bullet^5$ 1	2
			$\bullet^6$ determine equation of AB	$\bullet^6$ $y=x-3$	
	(c)		$\bullet^7$ find x or y coordinate	$\bullet^7$ $x=8$ or $y=5$	2
			$\bullet^8$ find remaining coordinate	$\bullet^8$ $y=5$ or $x=8$	
2.	(a)		**Method 1**	**Method 1**	2
			$\bullet^1$ know to use $x=1$ in synthetic division	$\bullet^1$ $\begin{array}{r|rrrr} 1 & 2 & -5 & 1 & 2 \\ \hline & 2 & & & \end{array}$	
			$\bullet^2$ complete division, interpret result and state conclusion	$\bullet^2$ $\begin{array}{r|rrrr} 1 & 2 & -5 & 1 & 2 \\ & & 2 & -3 & -2 \\ \hline & 2 & -3 & -2 & 0 \end{array}$ Remainder $=0$ $\therefore (x-1)$ is a factor	
			Method 2	**Method 2**	
			$\bullet^1$ know to substitute $x=1$	$\bullet^1$ $2(1)^3-5(1)^2+(1)+2$	
			$\bullet^2$ complete evaluation, interpret result and state conclusion	$\bullet^2$ $=0$ $\therefore (x-1)$ is a factor	
			Method 3	**Method 3**	
			$\bullet^1$ start long division and find leading term in quotient	$\bullet^1$ $(x-1)\overline{\smash{\big)}2x^3-5x^2+x+2}$ with quotient $2x^2$	
			$\bullet^2$ complete division, interpret result and state conclusion	$\bullet^2$ $(x-1)\overline{\smash{\big)}2x^3-5x^2+x+2}$ $\underline{2x^3-2x^2}$ $-3x^2+x$ $\underline{-3x^2+3x}$ $-2x+2$ $\underline{-2x+2}$ 0 quotient $2x^2-3x-2$ remainder $=0$ $\therefore (x-1)$ is a factor	

Question			Generic Scheme	Illustrative Scheme	Max mark
2.	(b)		$\bullet^3$ state quadratic factor	$\bullet^3$ $2x^2 - 3x - 2$	3
			$\bullet^4$ find remaining factors	$\bullet^4$ $(2x+1)$ and $(x-2)$	
			$\bullet^5$ state solution	$\bullet^5$ $x = -\dfrac{1}{2}, 1, 2$	
3.			$\bullet^1$ substitute for y	$\bullet^1$ $(x-2)^2 + (3x-1)^2 = 25$ or $x^2 - 4x + 4 + (3x)^2 - 2(3x) + 1 = 25$	5
			$\bullet^2$ express in standard quadratic form	$\bullet^2$ $10x^2 - 10x - 20 = 0$	
			$\bullet^3$ factorise	$\bullet^3$ $10(x-2)(x+1) = 0$	
			$\bullet^4$ find x coordinates	$\bullet^4$ $x = 2$ $\qquad \bullet^5$ $x = -1$	
			$\bullet^5$ find y coordinates	$\bullet^5$ $y = 6$ $\qquad$ $y = -3$	
4.	(a)		**Method 1**	**Method 1**	3
			$\bullet^1$ identify common factor	$\bullet^1$ $3(x^2 + 8x \ldots\ldots$ stated or implied by $\bullet^2$	
			$\bullet^2$ complete the square	$\bullet^2$ $3(x+4)^2 \ldots\ldots$	
			$\bullet^3$ process for c and write in required form	$\bullet^3$ $3(x+4)^2 + 2$	
			Method 2	**Method 2**	
			$\bullet^1$ expand completed square form	$\bullet^1$ $ax^2 + 2abx + ab^2 + c$	
			$\bullet^2$ equate coefficients	$\bullet^2$ $a = 3, \ 2ab = 24, \ ab^2 + c = 50$	
			$\bullet^3$ process for b and c and write in required form	$\bullet^3$ $3(x+4)^2 + 2$	
	(b)		$\bullet^4$ differentiate two terms	$\bullet^4$ $3x^2 + 24x\ldots$	2
			$\bullet^5$ complete differentiation	$\bullet^5$ $\ldots + 50$	
	(c)		**Method 1**	**Method 1**	2
			$\bullet^6$ link with (a) and identify sign of $(x+4)^2$	$\bullet^6$ $f'(x) = 3(x+4)^2 + 2$ and $(x+4)^2 \geq 0 \ \forall \ x$	
			$\bullet^7$ communicate reason	$\bullet^7$ $\therefore 3(x+4)^2 + 2 > 0 \Rightarrow$ always strictly increasing	
			Method 2	**Method 2**	
			$\bullet^6$ identify minimum value of $f'(x)$	$\bullet^6$ eg minimum value $= 2$ or annotated sketch	
			$\bullet^7$ communicate reason	$\bullet^7$ $2 > 0 \therefore (f'(x) > 0) \Rightarrow$ always strictly increasing	

Question			Generic Scheme	Illustrative Scheme	Max mark
5.	(a)		•¹ identify pathway	•¹ $\overrightarrow{PR}+\overrightarrow{RQ}$ stated or implied by •²	**2**
			•² state $\overrightarrow{PQ}$	•² $-3\mathbf{i}-4\mathbf{j}+5\mathbf{k}$	
	(b)		•³ interpret ratio	•³ $\dfrac{2}{3}$ or $\dfrac{1}{3}$	**2**
			•⁴ identify pathway and demonstrate result	•⁴ $\overrightarrow{PR}+\dfrac{2}{3}\overrightarrow{RQ}$ or $\overrightarrow{PQ}+\dfrac{1}{3}\overrightarrow{QR}$ leading to $\mathbf{i}-\mathbf{j}+4\mathbf{k}$	
	(c)		**Method 1**	**Method 1**	**5**
			•⁵ evaluate $\overrightarrow{PQ}.\overrightarrow{PS}$	•⁵ $\overrightarrow{PQ}.\overrightarrow{PS}=21$	
			•⁶ evaluate $\left|\overrightarrow{PQ}\right|$	•⁶ $\left|\overrightarrow{PQ}\right|=\sqrt{50}$	
			•⁷ evaluate $\left|\overrightarrow{PS}\right|$	•⁷ $\left|\overrightarrow{PS}\right|=\sqrt{18}$	
			•⁸ use scalar product	•⁸ $\cos QPS=\dfrac{21}{\sqrt{50}\times\sqrt{18}}$	
			•⁹ calculate angle	•⁹ $45\cdot6°$ or $0\cdot795$ radians	
			Method 2	**Method 2**	
			•⁵ evaluate $\left|\overrightarrow{QS}\right|$	•⁵ $\left|\overrightarrow{QS}\right|=\sqrt{26}$	
			•⁶ evaluate $\left|\overrightarrow{PQ}\right|$	•⁶ $\left|\overrightarrow{PQ}\right|=\sqrt{50}$	
			•⁷ evaluate $\left|\overrightarrow{PS}\right|$	•⁷ $\left|\overrightarrow{PS}\right|=\sqrt{18}$	
			•⁸ use cosine rule	•⁸ $\cos QPS=\dfrac{\left(\sqrt{50}\right)^2+\left(\sqrt{18}\right)^2-\left(\sqrt{26}\right)^2}{2\times\sqrt{50}\times\sqrt{18}}$	
			•⁹ calculate angle	•⁹ $45\cdot6°$ or $0\cdot795$ radians	
6.			•¹ substitute appropriate double angle formula	•¹ $5\sin x-4=2\left(1-2\sin^2 x\right)$	**5**
			•² express in standard quadratic form	•² $4\sin^2 x+5\sin x-6=0$	
			•³ factorise	•³ $(4\sin x-3)(\sin x+2)$	
				•⁴ •⁵	
			•⁴ solve for $\sin x°$	•⁴ $\sin x=\dfrac{3}{4},$ $\sin x=-2$	
			•⁵ solve for x	•⁵ $x=0\cdot848,\ 2\cdot29,\ \cancel{\sin x=-2}$	

Question			Generic Scheme	Illustrative Scheme	Max mark
7.	(a)		$\bullet^1$ write in differentiable form	$\bullet^1$ $\ldots -2x^{\frac{3}{2}}$ **stated or implied**	4
			$\bullet^2$ differentiate one term	$\bullet^2$ $\dfrac{dy}{dx} = 6 \ldots$ or $\dfrac{dy}{dx} = \ldots -3x^{\frac{1}{2}} \ldots$	
			$\bullet^3$ complete differentiation and equate to zero	$\bullet^3$ $\ldots -3x^{\frac{1}{2}} = 0$ or $6 \ldots = 0$	
			$\bullet^4$ solve for x	$\bullet^4$ $x = 4$	
	(b)		$\bullet^5$ evaluate y at stationary point	$\bullet^5$ 8	3
			$\bullet^6$ consider value of y at end points	$\bullet^6$ 4 and 0	
			$\bullet^7$ state greatest and least values	$\bullet^7$ greatest 8, least 0 **stated explicitly**	
8.	(a)		$\bullet^1$ find expression for u_1	$\bullet^1$ $5k - 20$	2
			$\bullet^2$ find expression for u_2 and express in the correct form	$\bullet^2$ $u_2 = k(5k - 20) - 20$ leading to $u_2 = 5k^2 - 20k - 20$	
	(b)		$\bullet^3$ interpret information	$\bullet^3$ $5k^2 - 20k - 20 < 5$	4
			$\bullet^4$ express inequality in standard quadratic form	$\bullet^4$ $5k^2 - 20k - 25 < 0$	
			$\bullet^5$ determine zeros of quadratic expression	$\bullet^5$ $-1, 5$	
			$\bullet^6$ state range with justification	$\bullet^6$ $-1 < k < 5$ with eg sketch or table of signs	

Question			Generic Scheme	Illustrative Scheme	Max mark
9.			**Method 1**	**Method 1**	5
			$\bullet^1$ state linear equation	$\bullet^1$ $\log_2 y = \dfrac{1}{4}\log_2 x + 3$	
			$\bullet^2$ introduce logs	$\bullet^2$ $\log_2 y = \dfrac{1}{4}\log_2 x + 3\log_2 2$	
			$\bullet^3$ use laws of logs	$\bullet^3$ $\log_2 y = \log_2 x^{\frac{1}{4}} + \log_2 2^3$	
			$\bullet^4$ use laws of logs	$\bullet^4$ $\log_2 y = \log_2 2^3 x^{\frac{1}{4}}$	
			$\bullet^5$ state k and n	$\bullet^5$ $k = 8,\ n = \dfrac{1}{4}$ or $y = 8x^{\frac{1}{4}}$	
			Method 2	**Method 2**	
			$\bullet^1$ state linear equation	$\bullet^1$ $\log_2 y = \dfrac{1}{4}\log_2 x + 3$	
			$\bullet^2$ use laws of logs	$\bullet^2$ $\log_2 y = \log_2 x^{\frac{1}{4}} + 3$	
			$\bullet^3$ use laws of logs	$\bullet^3$ $\log_2 \dfrac{y}{x^{\frac{1}{4}}} = 3$	
			$\bullet^4$ use laws of logs	$\bullet^4$ $\dfrac{y}{x^{\frac{1}{4}}} = 2^3$	
			$\bullet^5$ state k and n	$\bullet^5$ $k = 8,\ n = \dfrac{1}{4}$ or $y = 8x^{\frac{1}{4}}$	
			Method 3	**Method 3**	
				The equations at $\bullet^1$, $\bullet^2$ and $\bullet^3$ must be stated explicitly.	
			$\bullet^1$ introduce logs to $y = kx^n$	$\bullet^1$ $\log_2 y = \log_2 kx^n$	
			$\bullet^2$ use laws of logs	$\bullet^2$ $\log_2 y = n\log_2 x + \log_2 k$	
			$\bullet^3$ interpret intercept	$\bullet^3$ $\log_2 k = 3$	
			$\bullet^4$ use laws of logs	$\bullet^4$ $k = 8$	
			$\bullet^5$ interpret gradient	$\bullet^5$ $n = \dfrac{1}{4}$	
			Method 4	**Method 4**	
			$\bullet^1$ interpret point on log graph	$\bullet^1$ $\log_2 x = -12$ **and** $\log_2 y = 0$	
			$\bullet^2$ convert from log to exp. form	$\bullet^2$ $x = 2^{-12}$ **and** $y = 2^0$	
			$\bullet^3$ interpret point and convert	$\bullet^3$ $\log_2 x = 0,\ \log_2 y = 3$ $x = 1,\ y = 2^3$	
			$\bullet^4$ substitute into $y = kx^n$ and evaluate k	$\bullet^4$ $2^3 = k \times 1^n \Rightarrow k = 8$	
			$\bullet^5$ substitute other point into $y = kx^n$ and evaluate n	$\bullet^5$ $2^0 = 2^3 \times 2^{-12n}$ $\Rightarrow 3 - 12n = 0$ $\Rightarrow n = \dfrac{1}{4}$	

Question			Generic Scheme	Illustrative Scheme	Max mark
10.	(a)		**Method 1**	**Method 1**	3
			$\bullet^1$ calculate m_{AB}	$\bullet^1$ $m_{AB} = \dfrac{3}{9} = \dfrac{1}{3}$	
			$\bullet^2$ calculate m_{BC}	$\bullet^2$ $m_{BC} = \dfrac{5}{15} = \dfrac{1}{3}$	
			$\bullet^3$ interpret result and state conclusion	$\bullet^3$ $\ldots \Rightarrow$ AB and BC are parallel (common direction), B is a common point, hence A, B and C are collinear.	
			Method 2	**Method 2**	
			$\bullet^1$ calculate an appropriate vector e.g. $\overrightarrow{AB}$	$\bullet^1$ $\overrightarrow{AB} = \begin{pmatrix} 9 \\ 3 \end{pmatrix}$	
			$\bullet^2$ calculate a second vector e.g. $\overrightarrow{BC}$ and compare	$\bullet^2$ $\overrightarrow{BC} = \begin{pmatrix} 15 \\ 5 \end{pmatrix} \therefore \overrightarrow{AB} = \dfrac{3}{5}\overrightarrow{BC}$	
			$\bullet^3$ interpret result and state conclusion	$\bullet^3$ $\ldots \Rightarrow$ AB and BC are parallel (common direction), B is a common point, hence A, B and C are collinear.	
			Method 3	**Method 3**	
			$\bullet^1$ calculate m_{AB}	$\bullet^1$ $m_{AB} = \dfrac{3}{9} = \dfrac{1}{3}$	
			$\bullet^2$ find equation of line and substitute point	$\bullet^2$ eg, $y - 1 = \dfrac{1}{3}(x - 2)$ leading to $6 - 1 = \dfrac{1}{3}(17 - 2)$	
			$\bullet^3$ communication	$\bullet^3$ since C lies on line A, B and C are collinear	
	(b)		$\bullet^4$ find radius	$\bullet^4$ $6\sqrt{10}$	4
			$\bullet^5$ determine an appropriate ratio	$\bullet^5$ e.g. 2:3 or $\dfrac{2}{5}$ (using B and C) or 3:5 or $\dfrac{8}{5}$ (using A and C)	
			$\bullet^6$ find centre	$\bullet^6$ $(8,3)$	
			$\bullet^7$ state equation of circle	$\bullet^7$ $(x-8)^2 + (y-3)^2 = 360$	

Question			Generic Scheme	Illustrative Scheme	Max mark
11.	(a)		**Method 1**	**Method 1**	3
			$\bullet^1$ substitute for $\sin 2x$	$\bullet^1$ $\dfrac{2\sin x \cos x}{2\cos x} - \sin x \cos^2 x$ stated explicitly as above or in a simplified form of the above	
			$\bullet^2$ simplify and factorise	$\bullet^2$ $\sin x\left(1 - \cos^2 x\right)$	
			$\bullet^3$ substitute for $1 - \cos^2 x$ and simplify	$\bullet^3$ $\sin x \times \sin^2 x$ leading to $\sin^3 x$	
			Method 2	**Method 2**	
			$\bullet^1$ substitute for $\sin 2x$	$\bullet^1$ $\dfrac{2\sin x \cos x}{2\cos x} - \sin x \cos^2 x$ stated explicitly as above or in a simplified form of the above	
			$\bullet^2$ simplify and substitute for $\cos^2 x$	$\bullet^2$ $\sin x - \sin x\left(1 - \sin^2 x\right)$	
			$\bullet^3$ expand and simplify	$\bullet^3$ $\sin x - \sin x + \sin^3 x$ leading to $\sin^3 x$	
	(b)		$\bullet^4$ know to differentiate $\sin^3 x$	$\bullet^4$ $\dfrac{d}{dx}(\sin^3 x)$	3
			$\bullet^5$ start to differentiate	$\bullet^5$ $3\sin^2 x....$	
			$\bullet^6$ complete differentiation	$\bullet^6$ $...\times \cos x$	

HIGHER MATHEMATICS
2018

Paper 1 (Non-Calculator)

Question			Generic Scheme	Illustrative Scheme	Max mark
1.			•¹ find mid-point of PQ •² find gradient of median •³ determine equation of median	•¹ $(1,2)$ •² 2 •³ $y = 2x$	3
2.			**Method 1** •¹ equate composite function to x •² write $g(g^{-1}(x))$ in terms of $g^{-1}(x)$ •³ state inverse function **Method 2** •¹ write as $y = \frac{1}{5}x - 4$ and start to rearrange •² express x in terms of y •³ state inverse function **Method 3** •¹ interchange variables •² express y in terms of x •³ state inverse function	**Method 1** •¹ $g\left(g^{-1}(x)\right) = x$ •² $\frac{1}{5}g^{-1}(x) - 4 = x$ •³ $g^{-1}(x) = 5(x+4)$ **Method 2** •¹ $y + 4 = \frac{1}{5}x$ •² eg $x = 5(y+4)$ or $x = \dfrac{(y+4)}{\frac{1}{5}}$ •³ $g^{-1}(x) = 5(x+4)$ **Method 3** •¹ $x = \frac{1}{5}y - 4$ •² eg $y = 5(x+4)$ or $y = \dfrac{(y+4)}{\frac{1}{5}}$ •³ $g^{-1}(x) = 5(x+4)$	3
3.			•¹ start to differentiate •² complete differentiation •³ evaluate derivative	•¹ $-3\sin 2x \ldots$ stated or implied by •² •² $\ldots \times 2$ •³ $-3\sqrt{3}$	3
4.			•¹ state centre of circle •² find gradient of radius •³ state gradient of tangent •⁴ state equation of tangent	•¹ $(6,3)$ •² -4 •³ $\frac{1}{4}$ •⁴ $y = \frac{1}{4}x - 7$	4
5.	(a)		•¹ state ratio explicitly	•¹ $4:1$	1
	(b)		•² state value of t	•² 8	1

Question			Generic Scheme	Illustrative Scheme	Max mark
6.			•^{1} apply $m\log_5 x = \log_5 x^m$	•^{1} $\log_5 8^{\frac{1}{3}}$	3
			•^{2} apply $\log_5 x - \log_5 y = \log_5 \dfrac{x}{y}$	•^{2} $\log_5\left(\dfrac{250}{8^{\frac{1}{3}}}\right)$	
			•^{3} evaluate log	•^{3} 3	
7.	(a)		•^{1} state coordinates of P	•^{1} $(0,5)$	1
	(b)		•^{2} differentiate	•^{2} $3x^2 - 6x + 2$	3
			•^{3} calculate gradient	•^{3} 2	
			•^{4} state equation of tangent	•^{4} $y = 2x + 5$	
	(c)		•^{5} set $y_{\text{line}} = y_{\text{curve}}$ and arrange in standard form	•^{5} $x^3 - 3x^2 = 0$	4
			•^{6} factorise	•^{6} $x^2(x-3)$	
			•^{7} state x-coordinate of Q	•^{7} 3	
			•^{8} calculate y-coordinate of Q	•^{8} 11	
8.			•^{1} determine the gradient of the line	•^{1} $m = \sqrt{3}$ or $\tan\theta = \sqrt{3}$	2
			•^{2} determine the angle	•^{2} $60°$ or $\dfrac{\pi}{3}$	
9.	(a)		•^{1} identify pathway	•^{1} $-\mathbf{t} + \mathbf{u}$	1
	(b)		•^{2} state an appropriate pathway	•^{2} eg $\dfrac{1}{2}\overrightarrow{BC} + \overrightarrow{CA} + \overrightarrow{AD}$ stated or implied by •^{3}	2
			•^{3} express pathway in terms of $\mathbf{t}$, $\mathbf{u}$ and $\mathbf{v}$	•^{3} $-\dfrac{1}{2}\mathbf{t} - \dfrac{1}{2}\mathbf{u} + \mathbf{v}$	
10.			•^{1} know to and integrate one term	•^{1} eg $2x^3 \ldots$	4
			•^{2} complete integration	•^{2} eg $\ldots -\dfrac{3}{2}x^2 + 4x + c$	
			•^{3} substitute for x and y	•^{3} $14 = 2(2)^3 - \dfrac{3}{2}(2)^2 + 4(2) + c$	
			•^{4} state equation	•^{4} $y = 2x^3 - \dfrac{3}{2}x^2 + 4x - 4$ stated explicitly	

Question			Generic Scheme	Illustrative Scheme	Max mark
11.	(a)		$\bullet^1$ curve reflected in x-axis and translated 1 unit vertically $\bullet^2$ accurate sketch	$\bullet^1$ a generally decreasing curve above the x-axis for $1<x<3$ $\bullet^2$ asymptote at $x=0$ **and** passing through $(3,0)$ **and** continuing to decrease for $x\geq 3$	2
	(b)		$\bullet^3$ set '$y=y$' $\bullet^4$ start to solve $\bullet^5$ state x coordinate	$\bullet^3$ $\log_3 x = 1 - \log_3 x$ $\bullet^4$ $\log_3 x = \dfrac{1}{2}$ or $\log_3 x^2 = 1$ $\bullet^5$ $\sqrt{3}$ or $3^{\frac{1}{2}}$	3
12.	(a)		$\bullet^1$ find components	$\bullet^1$ $\begin{pmatrix} 6 \\ -3 \\ 4+p \end{pmatrix}$	1
	(b)		$\bullet^2$ find an expression for magnitude $\bullet^3$ start to solve $\bullet^4$ find values of p	$\bullet^2$ $\sqrt{6^2 + (-3)^2 + (4+p)^2}$ $\bullet^3$ $45 + (4+p)^2 = 49 \Rightarrow (4+p)^2 = 4$ or $p^2 + 8p + 12 = 0$ $\bullet^4$ $p=-2$, $p=-6$	3
13.	(a)	(i)	$\bullet^1$ find the value of $\cos x$ $\bullet^2$ substitute into the formula for $\sin 2x$ $\bullet^3$ simplify	$\bullet^1$ $\dfrac{\sqrt{7}}{\sqrt{11}}$ stated or implied by $\bullet^2$ $\bullet^2$ $2 \times \dfrac{2}{\sqrt{11}} \times \dfrac{\sqrt{7}}{\sqrt{11}}$ $\bullet^3$ $\dfrac{4\sqrt{7}}{11}$	3
		(ii)	$\bullet^4$ evaluate $\cos 2x$	$\bullet^4$ $\dfrac{3}{11}$	1
	(b)		$\bullet^5$ expand using the addition formula $\bullet^6$ substitute in values $\bullet^7$ simplify	$\bullet^5$ $\sin 2x \cos x + \cos 2x \sin x$ stated or implied by $\bullet^6$ $\bullet^6$ $\dfrac{4\sqrt{7}}{11} \times \dfrac{\sqrt{7}}{\sqrt{11}} + \dfrac{3}{11} \times \dfrac{2}{\sqrt{11}}$ $\bullet^7$ $\dfrac{34}{11\sqrt{11}}$	3

Question	Generic Scheme	Illustrative Scheme	Max mark
14.	•¹ write in integrable form	•¹ $(2x+9)^{-\frac{2}{3}}$	5
	•² start to integrate	•² $\dfrac{(2x+9)^{\frac{1}{3}}}{\frac{1}{3}}\ldots$	
	•³ complete integration	•³ $\ldots\times\dfrac{1}{2}$	
	•⁴ process limits	•⁴ $\dfrac{3}{2}\left(2(9)+9\right)^{\frac{1}{3}}-\dfrac{3}{2}\left(2(-4)+9\right)^{\frac{1}{3}}$	
	•⁵ evaluate integral	•⁵ 3	
15.	•¹ root at $x=-4$ identifiable from graph	•¹	4
	•² stationary point touching x-axis when $x=2$ identifiable from graph	•²	
	•³ stationary point when $x=-2$ identifiable from graph	•³	
	•⁴ identify orientation of the cubic curve and $f'(0)>0$ identifiable from graph	•⁴	

Paper 2 (Calculator)

Question	Generic scheme	Illustrative scheme	Max mark
1.	•¹ state an integral to represent the shaded area	•¹ $\displaystyle\int_{-1}^{3}\left(3+2x-x^2\right)dx$	4
	•² integrate	•² $3x+\dfrac{2x^2}{2}-\dfrac{x^3}{3}$	
	•³ substitute limits	•³ $\left(3\times3+\dfrac{2\times3^2}{2}-\dfrac{3^3}{3}\right)$ $-\left(3\times(-1)+\dfrac{2\times(-1)^2}{2}-\dfrac{(-1)^3}{3}\right)$	
	•⁴ evaluate integral	•⁴ $\dfrac{32}{3}$ (units²)	

Question			Generic scheme	Illustrative scheme	Max mark
2.	(a)		$\bullet^1$ find $\mathbf{u}.\mathbf{v}$	$\bullet^1$ 24	1
	(b)		$\bullet^2$ find $\|\mathbf{u}\|$	$\bullet^2$ $\sqrt{26}$	4
			$\bullet^3$ find $\|\mathbf{v}\|$	$\bullet^3$ $\sqrt{138}$	
			$\bullet^4$ apply scalar product	$\bullet^4$ $\cos\theta° = \dfrac{24}{\sqrt{26}\sqrt{138}}$	
			$\bullet^5$ calculate angle	$\bullet^5$ $66\cdot38...°$ or $1\cdot16...$ radians	
3.			$\bullet^1$ differentiate	$\bullet^1$ $3x^2 - 7$	3
			$\bullet^2$ evaluate derivative at $x=2$	$\bullet^2$ 5	
			$\bullet^3$ interpret result	$\bullet^3$ $\left(f \text{ is}\right)$ increasing	
4.			**Method 1**	**Method 1**	3
			$\bullet^1$ identify common factor	$\bullet^1$ $-3\left(x^2 + 2x...\right.$ stated or implied by $\bullet^2$	
			$\bullet^2$ complete the square	$\bullet^2$ $-3\left(x+1\right)^2...$	
			$\bullet^3$ process for c	$\bullet^3$ $-3\left(x+1\right)^2 + 10$	
			Method 2	**Method 2**	
			$\bullet^1$ expand completed square form	$\bullet^1$ $ax^2 + 2abx + ab^2 + c$	
			$\bullet^2$ equate coefficients	$\bullet^2$ $a=-3$, $2ab=-6$ $ab^2+c=7$	
			$\bullet^3$ process for b and c and write in required form	$\bullet^3$ $-3\left(x+1\right)^2 + 10$	
5.	(a)		$\bullet^1$ find the midpoint of PQ	$\bullet^1$ $(6, 1)$	3
			$\bullet^2$ calculate m_{PQ} and state perp. gradient	$\bullet^2$ $-1 \Rightarrow m_{perp} = 1$	
			$\bullet^3$ find equation of L_1 in a simplified form	$\bullet^3$ $y = x - 5$	
	(b)		$\bullet^4$ determine y coordinate	$\bullet^4$ 5	2
			$\bullet^5$ state x coordinate	$\bullet^5$ 10	
	(c)		$\bullet^6$ calculate radius of the circle	$\bullet^6$ $\sqrt{50}$ stated or implied by $\bullet^7$	2
			$\bullet^7$ state equation of the circle	$\bullet^7$ $\left(x-10\right)^2 + \left(y-5\right)^2 = 50$	

Question			Generic scheme	Illustrative scheme	Max mark
6.	(a)	(i)	$\bullet^1$ start composite process	$\bullet^1$ $f(2x)$	2
			$\bullet^2$ substitute into expression	$\bullet^2$ $3+\cos 2x$	
		(ii)	$\bullet^3$ state second composite	$\bullet^3$ $2(3+\cos x)$	1
	(b)		$\bullet^4$ equate expressions from (a)	$\bullet^4$ $3+\cos 2x = 2(3+\cos x)$	6
			$\bullet^5$ substitute for $\cos 2x$ in equation	$\bullet^5$ $3+2\cos^2 x - 1 = 2(3+\cos x)$	
			$\bullet^6$ arrange in standard quadratic form	$\bullet^6$ $2\cos^2 x - 2\cos x - 4 = 0$	
			$\bullet^7$ factorise	$\bullet^7$ $2(\cos x - 2)(\cos x + 1)$	
			$\bullet^8$ solve for $\cos x$	$\bullet^8$ $\cos x = 2 \qquad \bullet^9 \cos x = -1$	
			$\bullet^9$ solve for x	$\bullet^9$ $\cos x = 2 \qquad x = \pi$ or eg 'no solution'	
7.	(a)	(i)	$\bullet^1$ use '2' in synthetic division or in evaluation of cubic	$\bullet^1$ $2\,\lfloor\,2\ {-}3\ {-}3\ \ 2$ $\qquad\ 2$ or $2\times(2)^3 - 3(2)^2 - 3\times(2)+2$	2
			$\bullet^2$ complete division/evaluation and interpret result	$\bullet^2$ $2\,\lfloor\,2\ \ {-}3\ \ {-}3\ \ \ 2$ $\qquad\quad\ 4\ \ \ 2\ \ {-}2$ $\qquad\ 2\ \ \ 1\ \ {-}1\,\lfloor\ 0$ Remainder $= 0 \therefore (x-2)$ is a factor or $f(2)=0 \therefore (x-2)$ is a factor	
		(ii)	$\bullet^3$ state quadratic factor	$\bullet^3$ $2x^2 + x - 1$	2
			$\bullet^4$ complete factorisation	$\bullet^4$ $(x-2)(2x-1)(x+1)$ stated explicitly	
	(b)		$\bullet^5$ demonstrate result	$\bullet^5$ $u_6 = a(2a-3)-1 = 2a^2 - 3a - 1$ leading to $u_7 = a(2a^2 - 3a - 1) - 1$ $\qquad\qquad = 2a^3 - 3a^2 - a - 1$	1
	(c)	(i)	$\bullet^6$ equate u_5 and u_7 **and** arrange in standard form	$\bullet^6$ $2a^3 - 3a^2 - 3a + 2 = 0$	3
			$\bullet^7$ solve cubic	$\bullet^7$ $a = 2, \ a = \dfrac{1}{2}, \ a = -1$	
			$\bullet^8$ discard invalid solutions for a	$\bullet^8$ $a = \dfrac{1}{2}$	
		(ii)	$\bullet^9$ calculate limit	$\bullet^9$ -2	1

Question			Generic scheme	Illustrative scheme	Max mark
8.	(a)		•¹ use compound angle formula	•¹ $k\cos x° \cos a° + k\sin x° \sin a°$ stated explicitly	4
			•² compare coefficients	•² $k\cos a° = 2$ and $k\sin a° = -1$ stated explicitly	
			•³ process for k	•³ $k = \sqrt{5}$	
			•⁴ process for a and express in required form	•⁴ $\sqrt{5}\cos(x - 333\cdot4\ldots)°$	
	(b)	(i)	•⁵ state minimum value	•⁵ $-3\sqrt{5}$ or $-\sqrt{45}$	1
		(ii)	**Method 1**	**Method 1**	2
			•⁶ start to solve	•⁶ $x - 333\cdot4 = 180$ leading to $x = 513\cdot4$	
			•⁷ state value of x	•⁷ $x = 153\cdot4\ldots$	
			Method 2	**Method 2**	
			•⁶ start to solve	•⁶ $x - 333\cdot4 = -180$	
			•⁷ state value of x	•⁷ $x = 153\cdot4\ldots$	
9.			•¹ express P in differentiable form	•¹ $2x + 128x^{-1}$	6
			•² differentiate	•² $2 - \dfrac{128}{x^2}$	
			•³ equate expression for derivative to 0	•³ $2 - \dfrac{128}{x^2} = 0$	
			•⁴ process for x	•⁴ 8	
			•⁵ verify nature	•⁵ table of signs for a derivative ∴ minimum	
				or $P''(8) = \dfrac{1}{2} > 0$ ∴ minimum	
			•⁶ evaluate P	•⁶ $P = 32$ or min value $= 32$	

Question			Generic scheme	Illustrative scheme	Max mark
10.			$\bullet^1$ use the discriminant	$\bullet^1$ $(m-3)^2 - 4 \times 1 \times m$	4
			$\bullet^2$ identify roots of quadratic expression	$\bullet^2$ 1, 9	
			$\bullet^3$ apply condition	$\bullet^3$ $(m-3)^2 - 4 \times 1 \times m > 0$	
			$\bullet^4$ state range with justification	$\bullet^4$ $m < 1$, $m > 9$ with eg sketch or table of signs	
11.	(a)		$\bullet^1$ substitute for P and t	$\bullet^1$ $50 = 100\left(1 - e^{3k}\right)$	4
			$\bullet^2$ arrange equation in the form $A = e^{kt}$	$\bullet^2$ $0 \cdot 5 = e^{3k}$ or $-0 \cdot 5 = -e^{3k}$	
			$\bullet^3$ simplify	$\bullet^3$ $\ln 0 \cdot 5 = 3k$	
			$\bullet^4$ solve for k	$\bullet^4$ $k = -0 \cdot 231$	
	(b)		$\bullet^5$ evaluate P for $t = 5$	$\bullet^5$ 68·5	2
			$\bullet^6$ interpret result	$\bullet^6$ 31·5% still queueing	
12.	(a)	(i)	$\bullet^1$ write down coordinates of centre	$\bullet^1$ $(13, -4)$	1
		(ii)	$\bullet^2$ substitute coordinates and process for c	$\bullet^2$ $13^2 + (-4)^2 + 14 \times 13 - 22 \times (-4) \ldots$ leading to $c = -455$	1
	(b)	(i)	$\bullet^3$ calculate two key distances	$\bullet^3$ two from $r_2 = 25$, $r_1 = 10$ and $r_2 - r_1 = 15$	2
			$\bullet^4$ state ratio	$\bullet^4$ $3:2$ or $2:3$	
		(ii)	$\bullet^5$ identify centre of C_2	$\bullet^5$ $(-7, 11)$ or $\begin{pmatrix} -7 \\ 11 \end{pmatrix}$	2
			$\bullet^6$ state coordinates of P	$\bullet^6$ $(5, 2)$	
	(c)		$\bullet^7$ state equation	$\bullet^7$ $(x-5)^2 + (y-2)^2 = 1600$ or $x^2 + y^2 - 10x - 4y - 1571 = 0$	1

Paper 1 (Non-Calculator)

Question			Generic Scheme	Illustrative Scheme	Max mark
1.			$\bullet^1$ differentiate	$\bullet^1$ $2x-4$	4
			$\bullet^2$ calculate gradient	$\bullet^2$ 6	
			$\bullet^3$ find the value of y	$\bullet^3$ 12	
			$\bullet^4$ find equation of tangent	$\bullet^4$ $y=6x-18$	
2.			$\bullet^1$ find the centre	$\bullet^1$ $(-3,4)$	3
			$\bullet^2$ calculate the radius	$\bullet^2$ $\sqrt{17}$	
			$\bullet^3$ state equation of circle	$\bullet^3$ $(x+3)^2+(y-4)^2=17$ or equivalent	
3.	(a)		$\bullet^1$ find gradient l_1	$\bullet^1$ $\dfrac{1}{\sqrt{3}}$	2
			$\bullet^2$ state gradient l_2	$\bullet^2$ $-\sqrt{3}$	
	(b)		$\bullet^3$ using $m=\tan\theta$	$\bullet^3$ $\tan\theta=-\sqrt{3}$	2
			$\bullet^4$ calculating angle	$\bullet^4$ $\theta=\dfrac{2\pi}{3}$ or $120°$	
4.			$\bullet^1$ complete integration	$\bullet^1$ $-\dfrac{1}{6}x^{-1}$	3
			$\bullet^2$ substitute limits	$\bullet^2$ $\left(-\dfrac{1}{6\times2}\right)-\left(-\dfrac{1}{6\times1}\right)$	
			$\bullet^3$ evaluate	$\bullet^3$ $\dfrac{1}{12}$	
5.			$\bullet^1$ find $\overrightarrow{CD}$	$\bullet^1$ $\begin{pmatrix} x-4 \\ -3 \\ -1 \end{pmatrix}$	4
			$\bullet^2$ find $\overrightarrow{AB}$	$\bullet^2$ $\begin{pmatrix} 5 \\ -10 \\ -5 \end{pmatrix}$	
			$\bullet^3$ equate scalar product to zero	$\bullet^3$ $5(x-4)+(-10)(-3)+(-5)(-1)=0$	
			$\bullet^4$ calculate value of x	$\bullet^4$ $x=-3$	

Question			Generic Scheme	Illustrative Scheme	Max mark
6.			$\bullet^1$ substitute into discriminant	$\bullet^1$ $(p+1)^2 - 4\times1\times9$	4
			$\bullet^2$ apply condition for no real roots	$\bullet^2$ $...<0$	
			$\bullet^3$ determine zeroes of quadratic expression	$\bullet^3$ $-7, 5$	
			$\bullet^4$ state range with justification	$\bullet^4$ $-7<p<5$ with eg sketch or table of signs	
7.			$\bullet^1$ substitute for y in equation of circle	$\bullet^1$ $x^2+(3x-5)^2+2x-4(3x-5)-5=0$	5
			$\bullet^2$ express in standard quadratic form	$\bullet^2$ $10x^2-40x+40=0$	
			$\bullet^3$ demonstrate tangency	$\bullet^3$ $10(x-2)^2=0$ only one solution implies tangency	
			$\bullet^4$ find x-coordinate	$\bullet^4$ $x=2$	
			$\bullet^5$ find y-coordinate	$\bullet^5$ $y=1$	
8.	(a)		$\bullet^1$ use appropriate strategy	$\bullet^1$ $(1)^3-4(1)^2+a(1)+b=0$	5
			$\bullet^2$ obtain an expression for a and b	$\bullet^2$ $a+b=3$	
			$\bullet^3$ obtain a second expression for a and b	$\bullet^3$ $2a+b=-4$	
			$\bullet^4$ find the value of a or b	$\bullet^4$ $a=-7$ or $b=10$	
			$\bullet^5$ find the second value	$\bullet^5$ $b=10$ or $a=-7$	
	(b)		$\bullet^6$ obtain quadratic factor	$\bullet^6$ $\left(x^2-3x-10\right)$	3
			$\bullet^7$ complete factorisation	$\bullet^7$ $(x-1)(x-5)(x+2)$	
			$\bullet^8$ state solutions	$\bullet^8$ $x=1,\ x=5,\ x=-2$	
9.	(a)		$\bullet^1$ interpret information	$\bullet^1$ $13=28m+6$	2
			$\bullet^2$ solve to find m	$\bullet^2$ $m=\dfrac{1}{4}$	
	(b)	(i)	$\bullet^3$ state condition	$\bullet^3$ a limit exists as $-1<\dfrac{1}{4}<1$	1
		(ii)	$\bullet^4$ know how to calculate limit	$\bullet^4$ $L=\dfrac{1}{4}L+6$	2
			$\bullet^5$ calculate limit	$\bullet^5$ $L=8$	

Question			Generic Scheme	Illustrative Scheme	Max mark
10.	(a)		$\bullet^1$ state value	$\bullet^1$ 2	1
	(b)		$\bullet^1$ use laws of logarithms	$\bullet^1$ $\log_4 x(x-6)$	5
			$\bullet^2$ link to part (a)	$\bullet^2$ $\log_4 x(x-6)=2$	
			$\bullet^3$ use laws of logarithms	$\bullet^3$ $x(x-6)=4^2$	
			$\bullet^4$ write in standard quadratic form	$\bullet^4$ $x^2-6x-16=0$	
			$\bullet^5$ solve for x and identify appropriate solution	$\bullet^5$ 8	
11.			$\bullet^1$ start to differentiate	$\bullet^1$ $3\times 4\sin^2 x\ldots$	3
			$\bullet^2$ complete differentiation	$\bullet^2$ $\ldots \times \cos x$	
			$\bullet^3$ evaluate derivative	$\bullet^3$ $\dfrac{-3\sqrt{3}}{2}$	
12.			$\bullet^1$ calculate lengths AC and AD	$\bullet^1$ $AC=\sqrt{17}$ and $AD=5$ stated or implied by $\bullet^3$	5
			$\bullet^2$ select appropriate formula and express in terms of p and q	$\bullet^2$ $\cos q \cos p + \sin q \sin p$ stated or implied by $\bullet^4$	
			$\bullet^3$ calculate two of $\cos p,\ \cos q,\ \sin p,\ \sin q$	$\bullet^3$ $\cos p = \dfrac{4}{\sqrt{17}}$, $\cos q = \dfrac{4}{5}$ $\sin p = \dfrac{1}{\sqrt{17}}$, $\sin q = \dfrac{3}{5}$	
			$\bullet^4$ calculate other two and substitute into formula	$\bullet^4$ $\dfrac{4}{5}\times\dfrac{4}{\sqrt{17}}+\dfrac{3}{5}\times\dfrac{1}{\sqrt{17}}$	
			$\bullet^5$ arrange into required form	$\bullet^5$ $\dfrac{19}{5\sqrt{17}}\times\dfrac{\sqrt{17}}{\sqrt{17}}=\dfrac{19\sqrt{17}}{85}$ or $\dfrac{19}{5\sqrt{17}}=\dfrac{19\sqrt{17}}{5\times 17}=\dfrac{19\sqrt{17}}{85}$	
13.			$\bullet^1$ know to and start to integrate	$\bullet^1$ eg $y=\dfrac{4}{2}x^2\ldots$	4
			$\bullet^2$ complete integration	$\bullet^2$ $y=\dfrac{4}{2}x^2-\dfrac{6}{3}x^3+c$	
			$\bullet^3$ substitute for x and y	$\bullet^3$ $9=2(-1)^2-2(-1)^3+c$	
			$\bullet^4$ state expression for y	$\bullet^4$ $y=2x^2-2x^3+5$	

Question			Generic Scheme	Illustrative Scheme	Max mark
14.	(a)		$\bullet^1$ use double angle formula	**Method 1:** Using factorisation $\bullet^1$ $2\cos^2 x° - 1...$ **stated or implied by** $\bullet^2$	5
			$\bullet^2$ express as a quadratic in $\cos x°$ $\bullet^3$ start to solve	$\bullet^2$ $2\cos^2 x° - 3\cos x° + 1 = 0$ $\bullet^3$ $(2\cos x° - 1)(\cos x° - 1)$ = 0 must appear at either of these lines to gain $\bullet^2$	
				Method 2: Using quadratic formula $\bullet^1$ $2\cos^2 x° - 1...$ **stated or implied by** $\bullet^2$ $\bullet^2$ $2\cos^2 x° - 3\cos x° + 1 = 0$ **stated explicitly** $\bullet^3$ $\dfrac{-(-3) \pm \sqrt{(-3)^2 - 4 \times 2 \times 1}}{2 \times 2}$ **In both methods:**	
			$\bullet^4$ reduce to equations in $\cos x°$ only $\bullet^5$ process solutions in given domain	$\bullet^4$ $\cos x° = \dfrac{1}{2}$ and $\cos x° = 1$ $\bullet^5$ $0, 60, 300$ Candidates who include 360 lose $\bullet^5$. **or** $\bullet^4$ $\cos x = 1$ and $x = 0$ $\bullet^5$ $\cos x° = \dfrac{1}{2}$ and $x = 60$ or 300 Candidates who include 360 lose $\bullet^5$.	
	(b)		$\bullet^6$ interpret relationship with (a) $\bullet^7$ state valid values	$\bullet^6$ $2x = 0$ and 60 and 300 $\bullet^7$ 0, 30, 150, 180, 210 and 330	2
15.	(a)		$\bullet^1$ interpret notation $\bullet^2$ complete process	$\bullet^1$ $g(x^3 - 1)$ $\bullet^2$ $3x^3 - 2$	2
	(b)		$\bullet^3$ start to rearrange for x $\bullet^4$ rearrange $\bullet^5$ state expression for $h(x)$	$\bullet^3$ $3x^3 = y + 2$ $\bullet^4$ $x = \sqrt[3]{\dfrac{y+2}{3}}$ $\bullet^5$ $h(x) = \sqrt[3]{\dfrac{x+2}{3}}$	3

Paper 2 (Calculator)

Question			Generic Scheme	Illustrative Scheme	Max mark
1.	(a)		•¹ calculate gradient of AB	•¹ $m_{AB} = -3$	3
			•² use property of perpendicular lines	•² $m_{alt} = \frac{1}{3}$	
			•³ determine equation of altitude	•³ $x - 3y = 4$	
	(b)		•⁴ calculate midpoint of AC	•⁴ $(4,5)$	3
			•⁵ calculate gradient of median	•⁵ $m_{BM} = 2$	
			•⁶ determine equation of median	•⁶ $y = 2x - 3$	
	(c)		•⁷ find x or y coordinate	•⁷ $x = 1$ or $y = -1$	2
			•⁸ find remaining coordinate	•⁸ $y = -1$ or $x = 1$	
2.			•¹ write in integrable form	•¹ $4x + x^{-2}$	4
			•² integrate one term	•² eg $\frac{4}{2}x^2 + \ldots$	
			•³ integrate other term	•³ $\ldots \frac{x^{-1}}{-1}$	
			•⁴ complete integration and simplify	•⁴ $2x^2 - x^{-1} + c$	
3.			•¹ value of a	•¹ 1	3
			•² value of b	•² -2	
			•³ calculate k	•³ -1	
4.	(a)		•¹ state components of $\overrightarrow{DB}$	•¹ $\begin{pmatrix} 2 \\ 2 \\ -6 \end{pmatrix}$	3
			•² state coordinates of M	•² $(2,0,0)$ stated or implied by •³	
			•³ state components of $\overrightarrow{DM}$	•³ $\begin{pmatrix} 0 \\ -2 \\ -6 \end{pmatrix}$	

Question			Generic Scheme	Illustrative Scheme	Max mark
4.	(b)		•^{4} evaluate $\overrightarrow{DB}.\overrightarrow{DM}$	•^{4} 32	5
			•^{5} evaluate $\left|\overrightarrow{DB}\right|$	•^{5} $\sqrt{44}$	
			•^{6} evaluate $\left|\overrightarrow{DM}\right|$	•^{6} $\sqrt{40}$	
			•^{7} use scalar product	•^{7} $\cos BDM = \dfrac{32}{\sqrt{44}\sqrt{40}}$	
			•^{8} calculate angle	•^{8} $40\cdot3°$ or 0.703 rads	
5.			•^{1} know to integrate and interpret limits	•^{1} $\displaystyle\int_{-3}^{0} \ldots dx$	5
			•^{2} use 'upper $-$ lower'	•^{2} $\displaystyle\int_{-3}^{0} (x^3 + 3x^2 + 2x + 3) - (2x + 3)\, dx$	
			•^{3} integrate	•^{3} $\dfrac{1}{4}x^4 + x^3$	
			•^{4} substitute limits	•^{4} $0 - \left(\dfrac{1}{4}(-3)^4 + (-3)^3\right)$	
			•^{5} evaluate area	•^{5} $\dfrac{27}{4}$ units^2	
6.	(a)		**Method 1**	**Method 1**	3
			•^{1} identify common factor	•^{1} $3(x^2 + 8x\ldots\ldots$ stated or implied by •^{2}	
			•^{2} complete the square	•^{2} $3(x+4)^2 \ldots\ldots$	
			•^{3} process for c and write in required form	•^{3} $3(x+4)^2 + 2$	
			Method 2	**Method 2**	3
			•^{1} expand completed square form	•^{1} $ax^2 + 2abx + ab^2 + c$	
			•^{2} equate coefficients	•^{2} $a = 3, \quad 2ab = 24, \quad ab^2 + c = 50$	
			•^{3} process for b and c and write in required form	•^{3} $3(x+4)^2 + 2$	
	(b)		•^{4} differentiate two terms	•^{4} $3x^2 + 24x\ldots.$	2
			•^{5} complete differentiation	•^{5} $\ldots. + 50$	

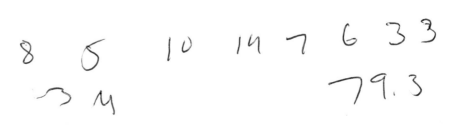

Question			Generic Scheme	Illustrative Scheme	Max mark
6.	(c)		**Method 1**	**Method 1**	**2**
			•^{6} link with (a) and identify sign of $(x+4)^2$	•^{6} $f'(x) = 3(x+4)^2 + 2$ and $(x+4)^2 \geq 0 \ \forall \, x$	
			•^{7} communicate reason	•^{7} $\therefore 3(x+4)^2 + 2 > 0 \Rightarrow$ always strictly increasing	
			Method 2	**Method 2**	**2**
			•^{6} identify minimum value of $f'(x)$	•^{6} eg minimum value $= 2$ or annotated sketch	
			•^{7} communicate reason	•^{7} $2 > 0 \therefore \left(f'(x) > 0\right) \Rightarrow$ always strictly increasing	
7.	(a)		•^{1} evidence of reflecting in x-axis	•^{1} reflection of graph in x-axis	**2**
			•^{2} vertical translation of 2 units identifiable from graph	•^{2} graph moves parallel to y-axis by 2 units upwards	
	(b)		•^{3} identify roots	•^{3} 0 **and** 2 only	**3**
			•^{4} interpret point of inflexion	•^{4} turning point at $(2,0)$	
			•^{5} complete cubic curve	•^{5} cubic passing through origin with negative gradient	

Question			Generic Scheme	Illustrative Scheme	Max mark
8.	(a)		$\bullet^1$ use compound angle formula	$\bullet^1$ $k\cos x\cos a - k\sin x\sin a$ **stated explicitly**	4
			$\bullet^2$ compare coefficients	$\bullet^2$ $k\cos a = 5, k\sin a = 2$ **stated explicitly**	
			$\bullet^3$ process for k	$\bullet^3$ $k = \sqrt{29}$	
			$\bullet^4$ process for a and express in required form	$\bullet^4$ $\sqrt{29}\cos(x+0\cdot38)$	
	(b)		$\bullet^5$ equate to 12 and simplify constant terms	$\bullet^5$ $5\cos x - 2\sin x = 2$ or $5\cos x - 2\sin x - 2 = 0$	4
			$\bullet^6$ use result of part (a) and rearrange	$\bullet^6$ $\cos(x+0\cdot3805...) = \dfrac{2}{\sqrt{29}}$	
			$\bullet^7$ solve for $x+a$	$\bullet^7$ 1·1902..., $\bullet^8$ 5·0928...	
			$\bullet^8$ solve for x	$\bullet^8$ 0·8097..., 4·712...	
9.	(a)		$\bullet^1$ equate volume to 100	$\bullet^1$ $V = \pi r^2 h = 100$	3
			$\bullet^2$ obtain an expression for h	$\bullet^2$ $h = \dfrac{100}{\pi r^2}$	
			$\bullet^3$ demonstrate result	$\bullet^3$ $A = \pi r^2 + 2\pi r^2 + 2\pi r \times \dfrac{100}{\pi r^2}$ leading to $A = \dfrac{200}{r} + 3\pi r^2$	
	(b)		$\bullet^4$ start to differentiate	$\bullet^4$ $A'(r) = 6\pi r...$	6
			$\bullet^5$ complete differentiation	$\bullet^5$ $A'(r) = 6\pi r - \dfrac{200}{r^2}$	
			$\bullet^6$ set derivative to zero	$\bullet^6$ $6\pi r - \dfrac{200}{r^2} = 0$	
			$\bullet^7$ obtain r	$\bullet^7$ $r = \sqrt[3]{\dfrac{100}{3\pi}}$ $(\approx 2\cdot20)$ metres	
			$\bullet^8$ verify nature of stationary point	$\bullet^8$ table of signs for a derivative when $r = 2\cdot1974...$	
			$\bullet^9$ interpret and communicate result	$\bullet^9$ minimum when $r \approx 2\cdot20$ (m) or $\bullet^8$ $A''(r) = 6\pi + \dfrac{400}{r^3}$ $\bullet^9$ $A''(2\cdot1974...) > 0$ ∴ minimum when $r \approx 2\cdot20$ (m)	

Question	Generic Scheme	Illustrative Scheme	Max mark
10.	•¹ start to integrate	•¹ $-\dfrac{1}{4}\cos\ldots$	6
	•² complete integration	•² $-\dfrac{1}{4}\cos\left(4x-\dfrac{\pi}{2}\right)$	
	•³ process limits	•³ $-\dfrac{1}{4}\cos\left(4a-\dfrac{\pi}{2}\right)+\dfrac{1}{4}\cos\left(\dfrac{4\pi}{8}-\dfrac{\pi}{2}\right)$	
	•⁴ simplify numeric term and equate to $\dfrac{1}{2}$	•⁴ $-\dfrac{1}{4}\cos\left(4a-\dfrac{\pi}{2}\right)+\dfrac{1}{4}=\dfrac{1}{2}$	
	•⁵ start to solve equation	•⁵ $\cos\left(4a-\dfrac{\pi}{2}\right)=-1$	
	•⁶ solve for a	•⁶ $a=\dfrac{3\pi}{8}$	
11.	**Method 1**	**Method 1**	3
	•¹ substitute for $\sin 2x$	•¹ $\dfrac{2\sin x\cos x}{2\cos x}-\sin x\cos^2 x$ stated explicitly as above or in a simplified form of the above	
	•² simplify and factorise	•² $\sin x\left(1-\cos^2 x\right)$	
	•³ substitute for $1-\cos^2 x$ and simplify	•³ $\sin x\times\sin^2 x$ leading to $\sin^3 x$	
	Method 2	**Method 2**	3
	•¹ substitute for $\sin 2x$	•¹ $\dfrac{2\sin x\cos x}{2\cos x}-\sin x\cos^2 x$ stated explicitly as above or in a simplified form of the above	
	•² simplify and substitute for $\cos^2 x$	•² $\sin x-\sin x\left(1-\sin^2 x\right)$	
	•³ expand and simplify	•³ $\sin x-\sin x+\sin^3 x$ leading to $\sin^3 x$	
12. (a)	**Method 1**	**Method 1**	3
	•¹ calculate m_{AB}	•¹ eg $m_{AB}=\dfrac{3}{9}=\dfrac{1}{3}$	
	•² calculate m_{BC}	•² eg $m_{BC}=\dfrac{5}{15}=\dfrac{1}{3}$	
	•³ interpret result and state conclusion	•³ $\ldots\Rightarrow$ AB and BC are parallel (common direction), B is a common point, hence A, B and C are collinear.	

Question			Generic Scheme	Illustrative Scheme	Max mark
12.	(a)		**Method 2**	**Method 2**	3
			$\bullet^1$ calculate an appropriate vector, eg AB	$\bullet^1$ eg $\overrightarrow{AB} = \begin{pmatrix} 9 \\ 3 \end{pmatrix}$	
			$\bullet^2$ calculate a second vector, eg $\overrightarrow{BC}$ and compare	$\bullet^2$ eg $\overrightarrow{BC} = \begin{pmatrix} 15 \\ 5 \end{pmatrix}$ $\therefore$ $\overrightarrow{AB} = \frac{3}{5}\overrightarrow{BC}$	
			$\bullet^3$ interpret result and state conclusion	$\bullet^3$ $\ldots \Rightarrow$ AB and BC are parallel (common direction), B is a common point, hence A, B and C are collinear.	
			Method 3	**Method 3**	3
			$\bullet^1$ calculate m_{AB}	$\bullet^1$ $m_{AB} = \frac{3}{9} = \frac{1}{3}$	
			$\bullet^2$ find equation of line and substitute point	$\bullet^2$ eg, $y - 1 = \frac{1}{3}(x - 2)$ leading to $6 - 1 = \frac{1}{3}(17 - 2)$	
			$\bullet^3$ communication	$\bullet^3$ since C lies on line A, B and C are collinear	
	(b)		$\bullet^4$ find radius	$\bullet^4$ $6\sqrt{10}$	4
			$\bullet^5$ determine an appropriate ratio	$\bullet^5$ eg 2:3 or $\frac{2}{5}$ (using B and C) or 3:5 or $\frac{8}{5}$ (using A and C)	
			$\bullet^6$ find centre	$\bullet^6$ $(8, 3)$	
			$\bullet^7$ state equation of circle	$\bullet^7$ $(x - 8)^2 + (y - 3)^2 = 360$	
13.	(a)		$\bullet^1$ interpret half-life	$\bullet^1$ $\frac{1}{2}P_0 = P_0 e^{-25k}$ **stated or implied by $\bullet^2$**	4
			$\bullet^2$ process equation	$\bullet^2$ $e^{-25k} = \frac{1}{2}$	
			$\bullet^3$ write in logarithmic form	$\bullet^3$ $\log_e \frac{1}{2} = -25k$	
			$\bullet^4$ process for k	$\bullet^4$ $k \approx 0.028$	
	(b)		$\bullet^5$ interpret equation	$\bullet^5$ $P_t = P_0 e^{-80 \times 0.028}$	3
			$\bullet^6$ process	$\bullet^6$ $P_t \approx 0.1065 P_0$	
			$\bullet^7$ state percentage decrease	$\bullet^7$ 89%	

Table of signs for a derivative — acceptable responses

x	-4^-	-4	-4^+
$\dfrac{dy}{dx}$ or $f'(x)$	$+$	0	$-$
Shape or slope	/	—	\

x	2^-	2	2^+
$\dfrac{dy}{dx}$ or $f'(x)$	$-$	0	$+$
Shape or slope	\	—	/

x	$\rightarrow$	-4	$\rightarrow$
$\dfrac{dy}{dx}$ or $f'(x)$	$+$	0	$-$
Shape or slope	/	—	\

x	$\rightarrow$	2	$\rightarrow$
$\dfrac{dy}{dx}$ or $f'(x)$	$-$	0	$+$
Shape or slope	\	—	/

Arrows are taken to mean "in the neighbourhood of".

x	a	-4	b	2	c
$\dfrac{dy}{dx}$ or $f'(x)$	$+$	0	$-$	0	$+$
Shape or slope	/	—	\	—	/

Where: $a < -4$, $-4 < b < 2$, $c > 2$.

Since the function is continuous "$-4 < b < 2$" is acceptable.

x	$\rightarrow$	-4	$\rightarrow$	2	$\rightarrow$
$\dfrac{dy}{dx}$ or $f'(x)$	$+$	0	$-$	0	$+$
Shape or slope	/	—	\	—	/

Since the function is continuous "$-4 \rightarrow 2$" is acceptable.